Understanding Organic Chemistry

Lecture & Workbook 5

Written by

Dr. Barbara A. van Kuiken

Harbour Books

Printed in the United States of America
1 2 3 4 5 6 7 8 9 10

Understanding Organic Chemistry: Lecture & Workbook 5
by Barbara A. van Kuiken

p. cm.
1. Organic Chemistry Laboratory Manual
2. Organic Chemistry Laboratory Techniques
3. Separation Chemistry

I. van Kuiken, Barbara A., 1960– II. Title.
ISBN 13: 978-1-7341842-3-5 (softcover)

Harbour Books
An imprint of
Mariner Media, Inc.
131 West 21st Street
Buena Vista, VA 24416
Tel: 540-264-0021
www.marinermedia.com

Contents

Chapter 14: **Understanding How to Analyze Structures of Products** **1**
(Part 1): Mass Spectrometry (MS)

Key Concepts 1

What You Need To Learn, Understand, and Apply 2

Chapter Preview 2

Lecture/Reading Notes 3

Learn to Analyze and Apply 15

Integrate Skills 18

Summary of Concepts and Analysis Methods 110

Chapter 15: **Understanding How to Analyze Structures of Products** **113**
(Part 2): Interpreting Infrared Spectroscopy (IR), Polarimetry,
and Ultraviolet-Visible Spectrophotometry (UV-Vis)

Key Concepts 113

What You Need To Learn, Understand, and Apply 114

Chapter Preview 114

Lecture/Reading Notes 115

Learn to Analyze and Apply 128

Integrate Skills 131

Summary of Concepts and Analysis Methods 218

Chapter 16: Understanding How to Analyze Structures of Products (Part 3): Nuclear Magnetic Resonance (NMR)   **221**

Key Concepts   221

What You Need To Learn, Understand, and Apply   222

Chapter Preview   223

Lecture/Reading Notes   224

Learn to Analyze and Apply   244

Integrate Skills   254

Summary of Concepts and Analysis Methods   355

Chapter 17: Understanding Organic Redox Reactions and Preparing for the ACS Exam   **357**

Key Concepts   357

What You Need To Learn, Understand, and Apply   357

Chapter Preview   358

Lecture/Reading Notes   359

Integrate Skills   381

Summary of Concepts and Analysis Methods   385

Daily Assignments

Date	Assignment	Due Date

Daily Assignments

Date	Assignment	Due Date

Chapter 14
Understanding How to Analyze Structures of Products (Part 1): Mass Spectrometry (MS)

Key Concepts

EI mass spectrometry creates radical cations by knocking an electron off each molecule. Non-bonded electrons, which are held by only one nucleus, are more loosely held than bond electrons, which are held by two nuclei. Therefore, non-bonded electrons are more likely to be removed than bond electrons. Because of the resulting instability, the molecule often fragments in characteristic ways, with carbocation stability often being an important consideration. The detector registers only those molecules or fragments that have a positive charge.

To simplify analysis of a mass spectrometry fragmentation pattern, focus on the following key features:

1. The M+ peak (which allows you to determine overall mass as well as the potential number of carbons). The mass of a CH_3 group is 15 g/mole. Each additional alkyl carbon group averages 14 g/mole.

2. Characteristic isotope patterns associated with the M+ peak (which allows you to determine the type of functional group present).

Characteristic patterns include:

Alkyl Bromides	Have an M+2 peak that is the same height as the M+ peak
Alky Chlorides	Have an M+2 peak that is 1/3 the height of the M+ peak
Alcohols	Are often missing the M+ peak, and have a characteristic M-18 peak caused by the loss of water. Alcohols often also have unusual m/z values, such as 31 or 45, caused by the loss of an R-OH group.
Amines	If an odd number of nitrogens are present, the M+ peak is an odd number
Aldehyde/Ketone	May undergo a McLafferty rearrangement if it has at least one γ hydrogen. Characterized by the loss of a group that has an m/z value one less than expected.

3. The ==base peak== value, as well as the difference between the base peak and M+ value, provide structural information since they indicate where fragmentation most often occurred.

What You Need To Learn, Understand, and Apply

1. The types of information a mass spectrum provides. (page 403)
2. The general theory of mass spectrometry. (page 403)
3. The definition for each of the following: m/z, M+, M+1, M+2, base peak. (pages 403–404)
4. The ability to determine the most common ways a radical cation fragments. (pages 405–407)
5. The ability to identify unique characteristics of fragmentation patterns for alkanes, alkyl halides, ethers, alcohols, amines, aldehydes, and ketones. (pages 407–412)
6. The skills needed to apply the material and avoid common errors. (page 412)

Chapter Preview

The general purpose of this chapter:

_____

_____

_____

_____

_____

(Keep this purpose in mind as you read the chapter to help you tie all the concepts together into one complete picture.)

Questions to Assess Understanding	Lecture/Reading After filling in your lecture note outline, fold the paper over so that only the assessment question is visible. Once you can consistently answer a given question correctly on your own, place an X by that question.
_____ What 3 pieces of information can be determined from a mass spectrum?	**Learning Objective 1: Know what types of information a mass spectrum provides. (page 403)** Mass spectrometry is used to determine: 1. _____ 2. _____ 3. _____
_____ What types of molecules are detected by a mass spectrometer?	**Learning Objective 2: Know the general theory of mass spectrometry. (page 403)** _____ _____ _____ _____ _____
_____ What is the definition of m/z, and what does it tell you? _____ What is the definition of an M+ value on a mass spectrum?	**Learning Objective 3: Know the definition for each of the following: m/z, M+, M+1, M+2, and base peak. (pages 403–404)** m/z _____ _____ _____

_____What is the definition of an M+1 value and of an M+2 value on a mass spectrum? _____ What is the definition of the base peak on a mass spectrum and what does it tell you?	M+ _____ M+1 _____ _____ M+2 _____ _____ Base peak _____ _____ _____

Example:

Mass Spectrum of N,N-Dimethylbutan-1-amine

Peaks: 58 (base peak, ~95 relative abundance), 101 (~15 relative abundance). Axes: Relative Abundance (0–100) vs m/z (0–200).

Learning Objective 4: Be able to determine the most common fragmentation patterns for cation radicals. (pages 405–407)

_____ What are the most likely places to lose an electron from a molecule and why?

The most likely place to lose an electron from a molecule and the reason why:

1. _____

_____

2. _____

_____

3. _____

The difference between heterolytic and homolytic (alpha) cleavage:

_____ What is the definition of heterolytic cleavage in mass spectrometry?

Heterolytic cleavage: _____

_____

_____

Example:

CH_3

H_3C——C——CH_3

:Y. +

Both a radical and a cation
(Measured by the detector)

A cation (Measured by the detector)

and

A radical but not a cation
(Not measured by the detector)

_____ What is the definition of homolytic cleavage in mass spectrometry?

Homolytic cleavage: _____

_____

_____

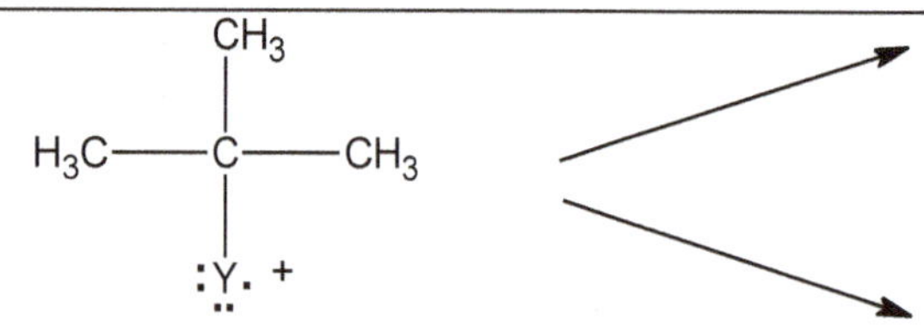

CH$_3$

H$_3$C———C———CH$_3$

:Y. +

A cation (Measured by the detector)

and

A radical but not a cation
(Not measured by the detector)

Both a radical and a cation
(Measured by the detector)

How to quickly determine what the most likely fragments are:

_____ What is one way to quickly determine the most likely fragments for a molecule subjected to mass spectrometric analysis?

_____

_____

_____

_____

Example:

(The M+ is 178)

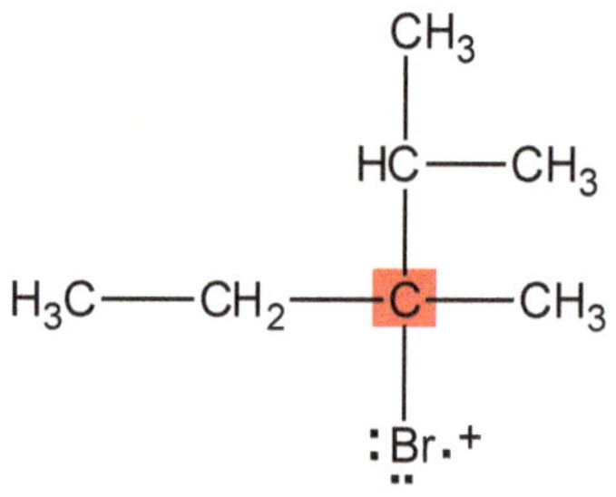

CH$_3$

HC———CH$_3$

H$_3$C———CH$_2$———C———CH$_3$

:Br.+

_____ If two alkyl fragments have the same stability, which is more likely to be removed?

If two alkyl fragments have similar stability, the one more likely to be removed is:

_____

Example:

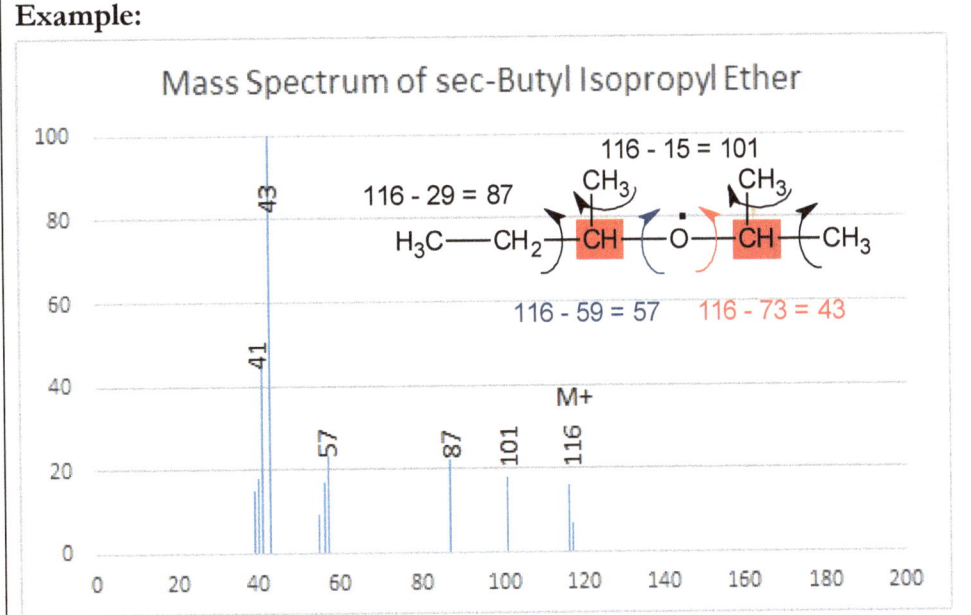

Learning Objective 5: Be able to identify unique characteristics of fragmentation patterns for alkanes, alkyl halides, ethers, alcohols, amines, aldehydes, and ketones. (pages 407–412)

Alkanes

_____ Where is an alkane most likely to fragment when an electron is removed from it during mass spectrometry?

The most likely place for an alkane to fragment _____

_____

Example:

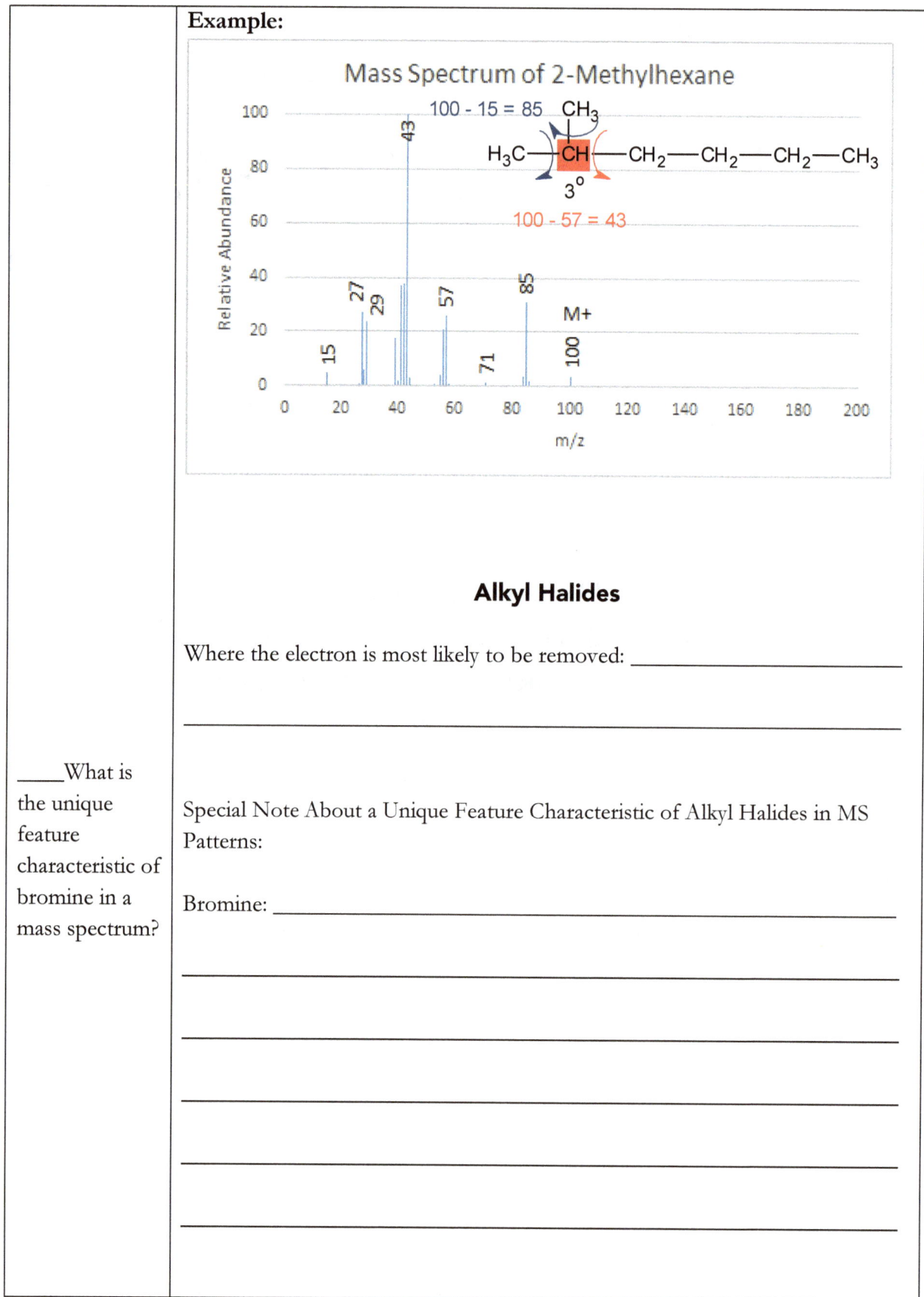

Alkyl Halides

Where the electron is most likely to be removed: _____

_____

_____What is the unique feature characteristic of bromine in a mass spectrum?

Special Note About a Unique Feature Characteristic of Alkyl Halides in MS Patterns:

Bromine: _____

_____

_____

_____

_____

_____

Example:

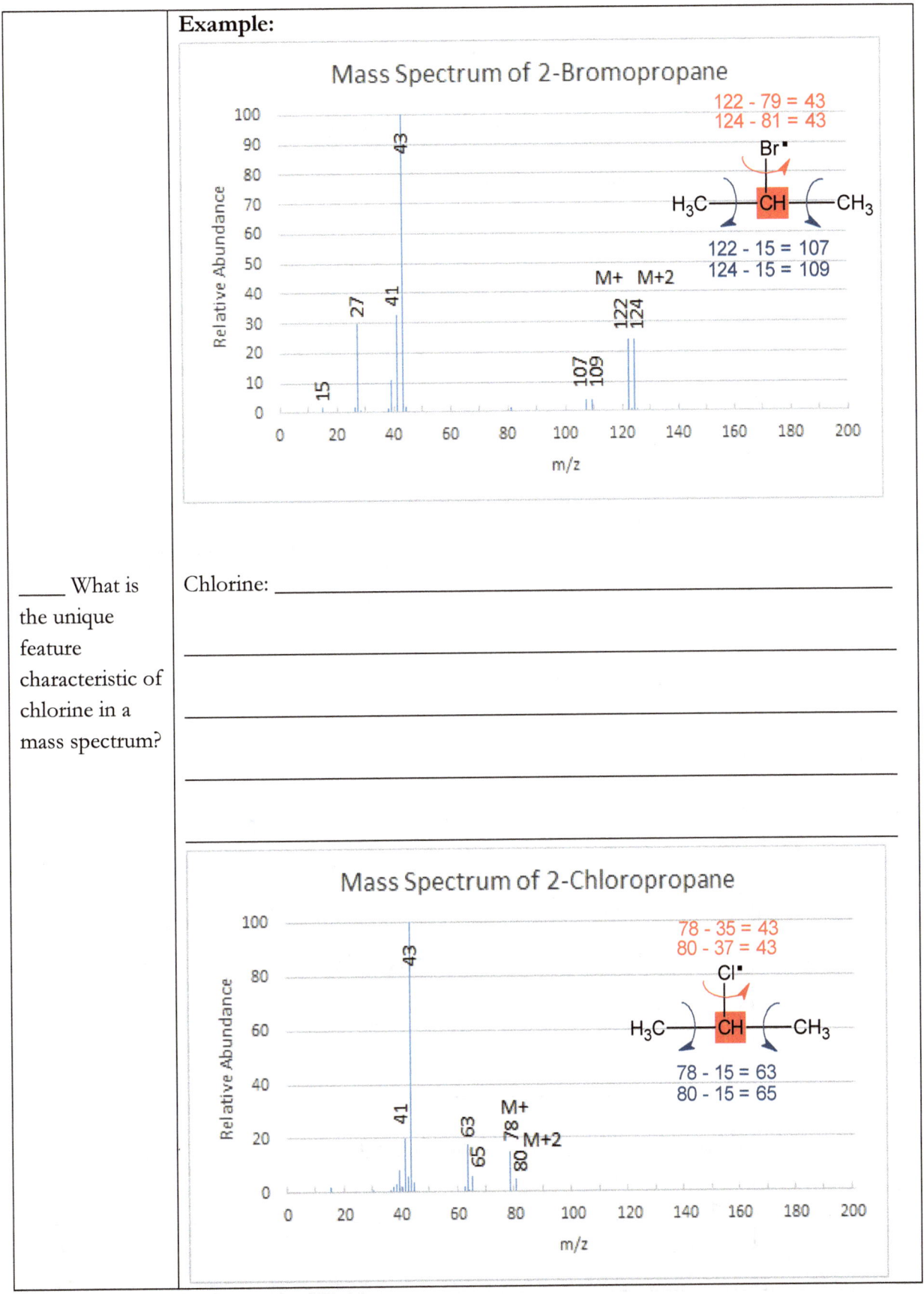

_____ What is the unique feature characteristic of chlorine in a mass spectrum?

Chlorine: _____

_____

_____

_____

_____ If an ether is not symmetrical, which alkyl group is more likely to be removed?	**Ethers** Where the electron is most likely to be removed: _____ _____ If the molecule is not symmetrical, the more likely place where alpha cleavage will occur: _____ **Example:** Mass Spectrum of sec-Butyl Isopropyl Ether 116 - 15 = 111 116 - 59 = 57 116 - 73 = 43 **Alcohols** Where the electron is most likely to be removed: _____ _____ Unique Characteristics of Alcohols in MS Patterns: 1. _____ _____

_____ What are three unique characteristics of alcohols that usually appear in a mass spectrum?

2. _____

_____

3. _____

_____

Example:

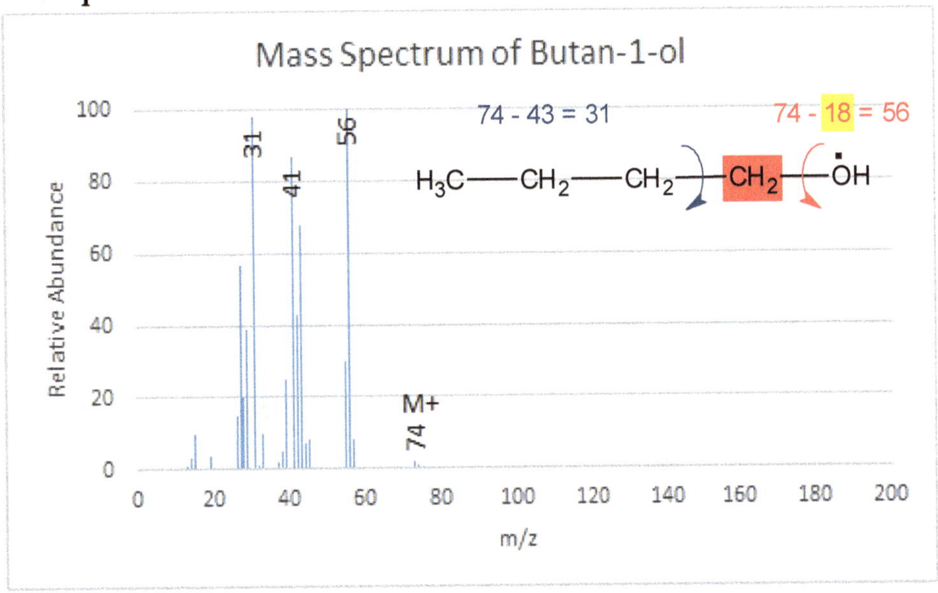

Mass Spectrum of Butan-1-ol

$74 - 43 = 31$ $74 - 18 = 56$

H_3C——CH_2——CH_2——CH_2——$\overset{..}{O}H$

Amines

Where the electron is most likely to be removed: _____

_____

_____ What unique feature is characteristic of amines in a mass spectrum?

A unique feature about the M+ value any amine that has an odd number of nitrogens:

_____

_____

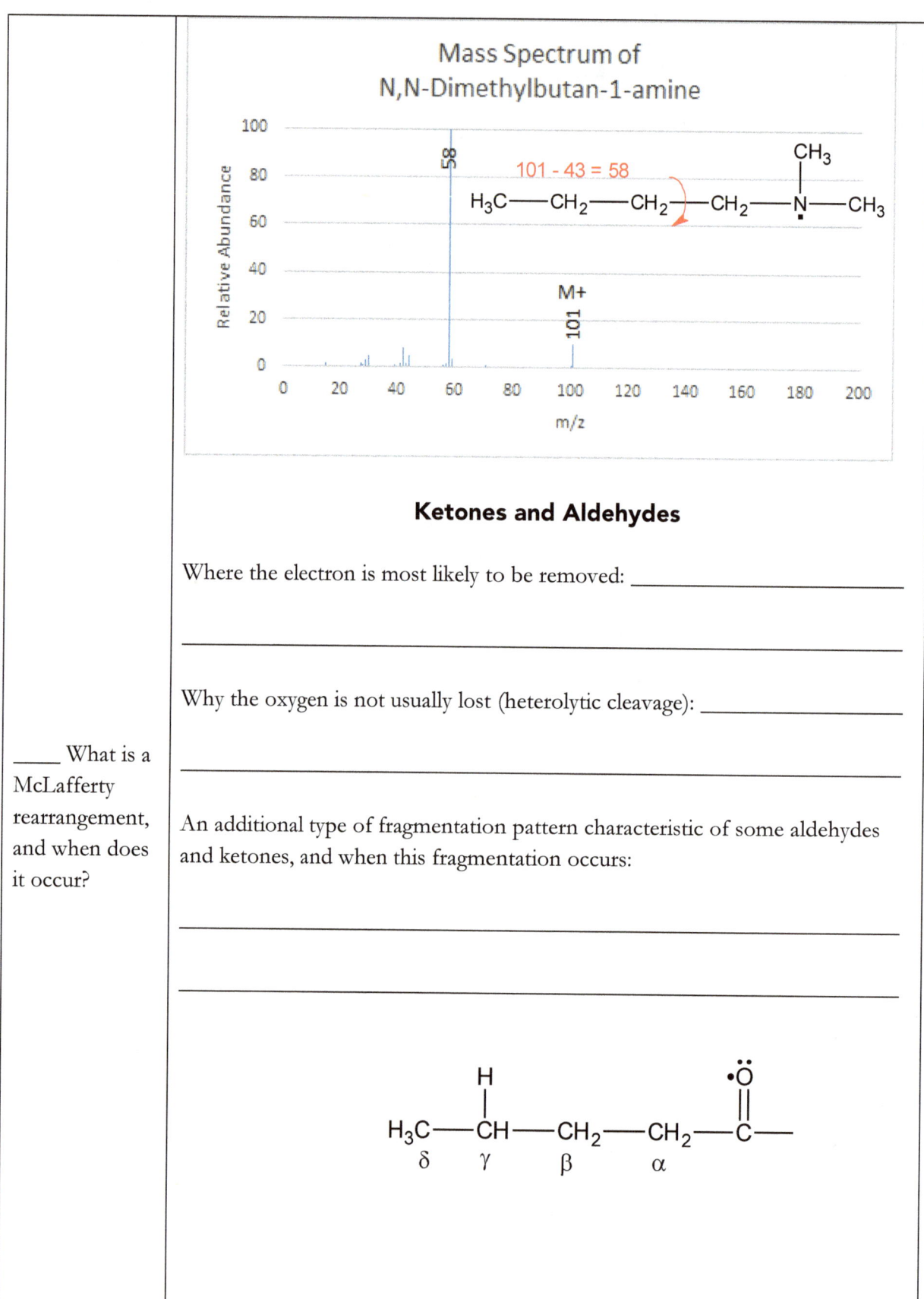

Ketones and Aldehydes

Where the electron is most likely to be removed: _____

_____

Why the oxygen is not usually lost (heterolytic cleavage): _____

_____

_____ What is a McLafferty rearrangement, and when does it occur?

An additional type of fragmentation pattern characteristic of some aldehydes and ketones, and when this fragmentation occurs:

_____

_____

Example:

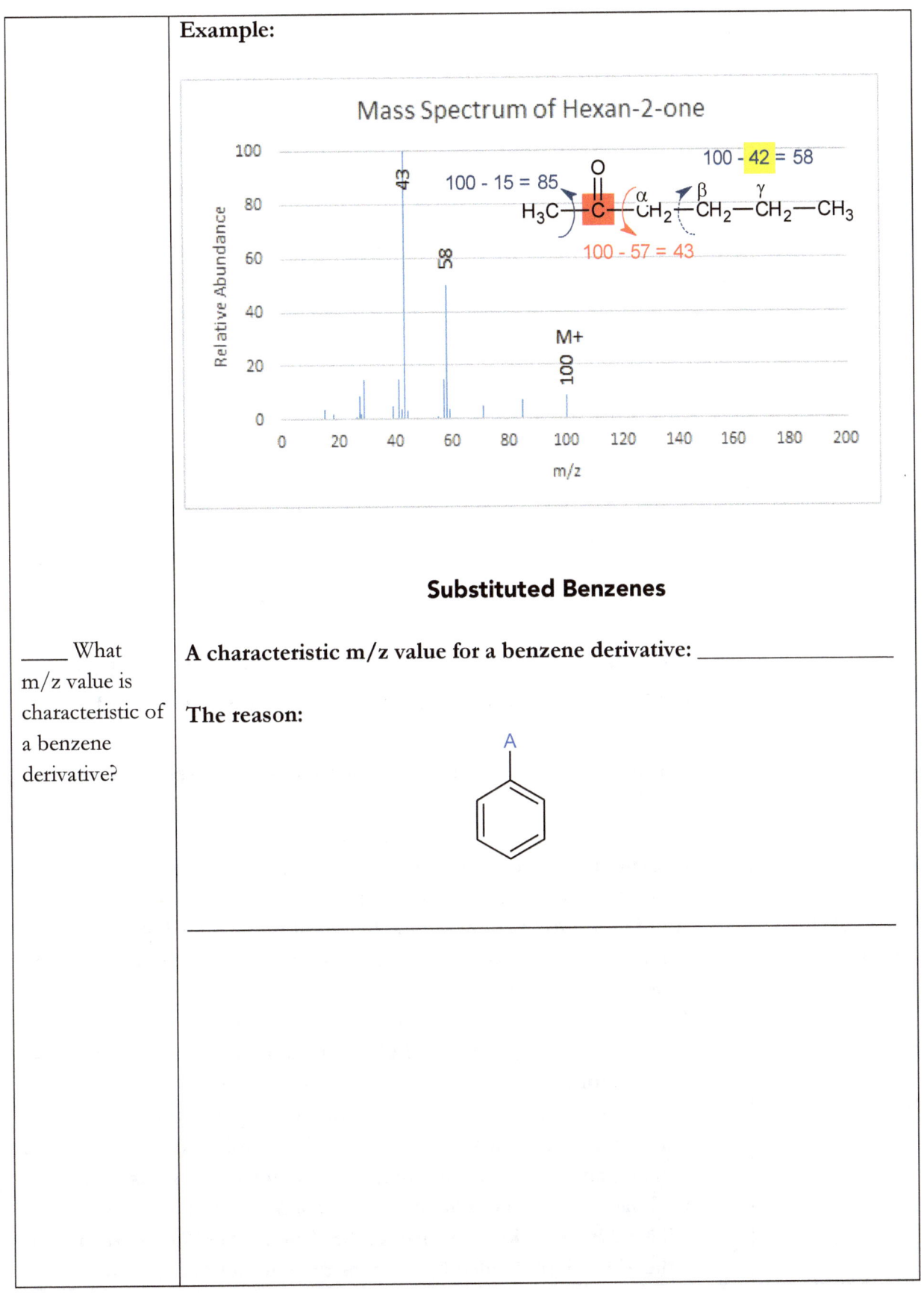

Substituted Benzenes

_____ What m/z value is characteristic of a benzene derivative?

A characteristic m/z value for a benzene derivative: _____

The reason:

Example:

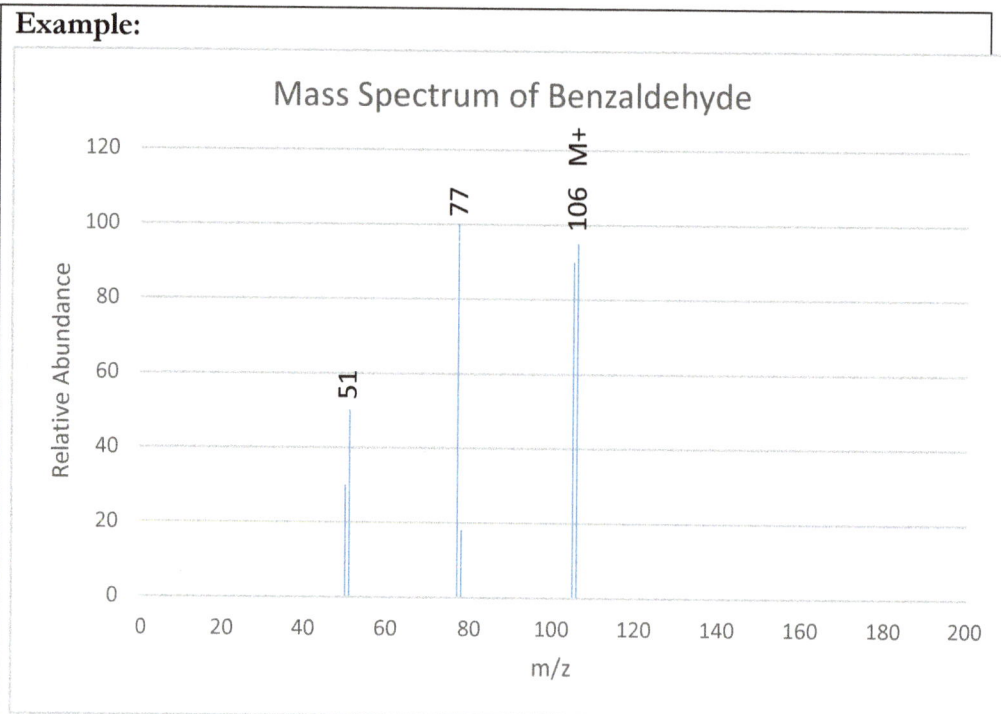

Mass Spectrum of Benzaldehyde

Learning Objective 6: Gain the skills needed to apply the material and avoid common errors. (page 412)

To determine the structure of a molecule from its mass spectrum:

STEP 1:
Start by gaining as much information as possible from the **M+** peak

An odd number = an odd number of nitrogens

An M+2 approximately 1/3 the height of the M+ peak = chlorine

An M + 2 approximately the same height as the M+ peak = bromine

A missing M+ peak = alcohol

If the functional group is known, subtract the molar mass of the functional group from the M+ value, and use the remaining mass to determine the number of carbons. (15 + 14n)

When possible, determine how many rings/double bonds the molecule has (each ring and each double bond uses up 2 hydrogens, and will, therefore, decrease the molar mass by 2.)

C-13 has a relative abundance of 1.1% compared to C-12. That means if M+1 information is available, the % of the M+1 height relative to the M+ height divided by 1.1 gives the number of carbons.

_____ What is the process for determining the structure of a molecule from its mass spectrum?

STEP 2:

When possible, determine the number of carbons/hydrogens represented by the BASE PEAK. Also, to determine what fragment is most easily lost, calculate the value for the M+ peak minus the base peak.

Other helpful information:

Benzene derivatives usually have a characteristic peak at an m/z value of 77.

Alcohols often have a peak with an m/z value of 31. Sometimes, they also have an M-18 peak.

STEP 3:

Sometimes it will be necessary to DRAW all potential configurations based on the known number of carbons and type of functional group. MATCH the major fragmentation patterns of each configuration against the known fragmentation pattern. Take into account both heterolytic and homolytic fragmentation.

Learn to Analyze and Apply

Determine the identity of the molecule that creates positively-charged fragments with the following m/z values:

31, 73 (base peak), 87, 98, 101, 116 (M+)

STEP 1:

Start by gaining as much information as possible from the M+ peak

An odd number = an odd number of nitrogens

An M+2 approximately 1/3 the height of the M+ peak = chlorine

An M + 2 approximately the same height as the M+ peak = bromine

A missing M+ peak = alcohol

If the functional group is known, subtract the molar mass of the functional group from the M+ value, and use the remaining mass to determine the number of carbons. (15 + 14n)

When possible, determine how many rings/double bonds the molecule has (each ring and each double bond uses up 2 hydrogens, and will, therefore, decrease the molar mass by 2.)

1C = 15
2C = 29
3C = 43
4C = 57
5C = 71
6C = 85
7C = 99

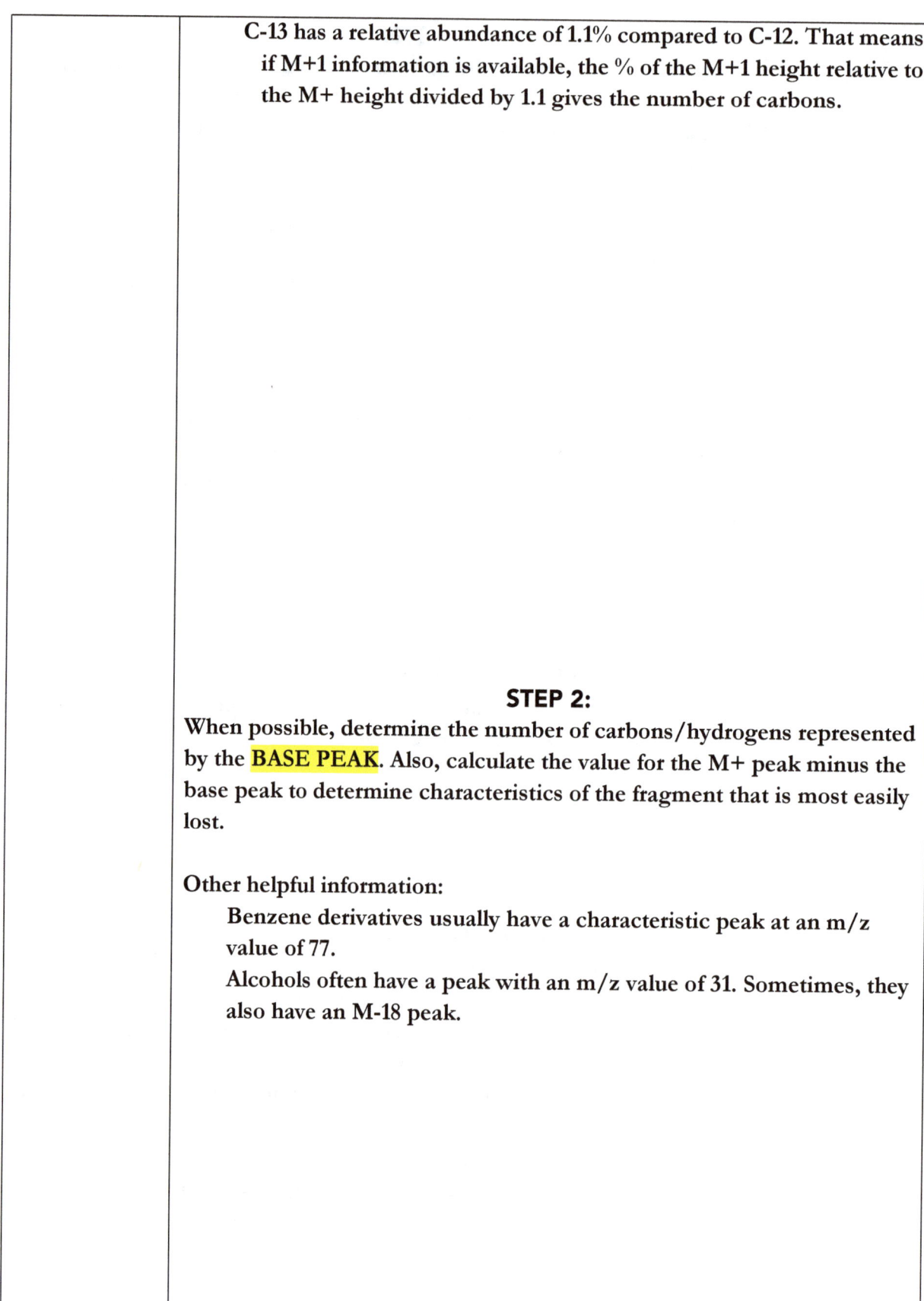

C-13 has a relative abundance of 1.1% compared to C-12. That means if M+1 information is available, the % of the M+1 height relative to the M+ height divided by 1.1 gives the number of carbons.

STEP 2:

When possible, determine the number of carbons/hydrogens represented by the BASE PEAK. Also, calculate the value for the M+ peak minus the base peak to determine characteristics of the fragment that is most easily lost.

Other helpful information:

Benzene derivatives usually have a characteristic peak at an m/z value of 77.

Alcohols often have a peak with an m/z value of 31. Sometimes, they also have an M-18 peak.

STEP 3:

Sometimes it will be necessary to <mark>DRAW</mark> all potential configurations based on the known number of carbons and type of functional group. <mark>MATCH</mark> the major fragmentation patterns of each configuration against the known fragmentation pattern. Take into account both heterolytic and homolytic fragmentation.

Integrate Skills

1. What is the m/z value for each heterolytic/homolytic fragment you would expect to be created during mass spec analysis of each of the following? (Remember that since mass spec fragmentation is complex, additional fragments may contribute significantly to the spectrum):

Alkanes

3,3-dimethylhexane

_____

3-ethylpentane

_____

Alkyl Halides

2-chloro-3-methylhexane

_____

3-bromohexane

_____

4-bromo-4-ethyl-3-methyloctane

_____

3-bromo-3-methylhexane

_____

4-bromo-4-ethyloctane

_____

3-bromo-3-ethylhexane

_____

Aldehydes and Ketones

3-methylbutanone

_____

hexan-2-one

_____

pentan-3-one

_____

propanal

_____

5-methylhexanal

_____

Ether

tert-butyl propyl ether

_____

sec-butyl isobutyl ether

_____

Alcohols

pentan-1-ol

_____

tert-butyl alcohol

_____

Amines

5-methylhexan-3-amine

_____

N-isopropylheptan-3-amine

_____

pentan-1-amine

_____

Substituted Benzenes

chlorobenzene

_____

aniline

_____

2. How might you be able to distinguish between each of the following based on mass spec data? (Write all major differences you could use.)

5,5-dimethylhexan-2-one and octan-2-one

_____

butan-1-ol and diethyl ether

_____

3-bromopentane and 3-chloropentane

_____

butanone and butanal

_____

tert-butylamine and *tert*-butyl alcohol

_____

1-chloropentane vs. 2-chloropentane

_____

3. For each question, use the data to determine whether the molecule is an alkyl chloride, alkyl bromide, amide, alcohol, ether, or carbonyl. Then, circle the specific structure within the selected category that most closely matches the data. (There is only one answer choice for each question.)

   a. Base peak = 57, M+ peak = 92, M+2 peak 1/3 height = 94, Additional peaks at 41, 77, and 79 (1/3 the height of the peak at 77)

       Choices:

       Alkyl chloride: *tert*-butyl chloride, *sec*-butyl chloride, butyl chloride,

       Alkyl bromide: *tert*-butyl bromide, *sec*-butyl bromide, butyl bromide,

       Amide: *tert*-butyl amine, *sec*-butyl amine, butyl amine,

Alcohol: *tert*-butyl alcohol, *sec*-butyl alcohol, butyl alcohol

Ether: *tert*-butyl ether, *sec*-butyl ether, butyl ether

Carbonyl: butanone, butanal

b.   Base peak = 57, M+ = 130, other prominent peak = 87

Choices:

Alkyl chloride: *tert*-butyl chloride, *sec*-butyl chloride, butyl chloride,

Alkyl bromide: *tert*-butyl bromide, *sec*-butyl bromide, butyl bromide,

Amide: *tert*-butyl amine, *sec*-butyl amine, butyl amine,

Alcohol: *tert*-butyl alcohol, *sec*-butyl alcohol, butyl alcohol

Ether: *tert*-butyl ether, *sec*-butyl ether, butyl ether

Carbonyl: butanone, butanal

c. Base peak = 44, M+ = 73, other prominent peak = 58

Choices:

Alkyl chloride: *tert*-butyl chloride, *sec*-butyl chloride, butyl chloride,

Alkyl bromide: *tert*-butyl bromide, *sec*-butyl bromide, butyl bromide,

Amide: *tert*-butyl amine, *sec*-butyl amine, butyl amine,

Alcohol: *tert*-butyl alcohol, *sec*-butyl alcohol, butyl alcohol

Ether: *tert*-butyl ether, *sec*-butyl ether, butyl ether

Carbonyl: butanone, butanal

d.   Base peak = 56, M+ = 74, other prominent peak = 31

Choices:

Alkyl chloride: *tert*-butyl chloride, *sec*-butyl chloride, butyl chloride,

Alkyl bromide: *tert*-butyl bromide, *sec*-butyl bromide, butyl bromide,

Amide: *tert*-butyl amine, *sec*-butyl amine, butyl amine,

Alcohol: *tert*-butyl alcohol, *sec*-butyl alcohol, butyl alcohol

Ether: *tert*-butyl ether, *sec*-butyl ether, butyl ether

Carbonyl: butanone, butanal

e. Base peak = 43, M+ = 72, other prominent peak = 57

Choices:

Alkyl chloride: *tert*-butyl chloride, *sec*-butyl chloride, butyl chloride,

Alkyl bromide: *tert*-butyl bromide, *sec*-butyl bromide, butyl bromide,

Amide: *tert*-butyl amine, *sec*-butyl amine, butyl amine,

Alcohol: *tert*-butyl alcohol, *sec*-butyl alcohol, butyl alcohol

Ether: *tert*-butyl ether, *sec*-butyl ether, butyl ether

Carbonyl: butanone, butanal

f. Base peak = 44, M+ = 72, other prominent peak = 57

Choices:

Alkyl chloride: *tert*-butyl chloride, *sec*-butyl chloride, butyl chloride,

Alkyl bromide: *tert*-butyl bromide, *sec*-butyl bromide, butyl bromide,

Amide: *tert*-butyl amine, *sec*-butyl amine, butyl amine,

Alcohol: *tert*-butyl alcohol, *sec*-butyl alcohol, butyl alcohol

Ether: *tert*-butyl ether, *sec*-butyl ether, butyl ether

Carbonyl: butanone, butanal

g. Base peak = 43, M+ peak = 100, Additional peaks at 58, 71, and 85

Carbonyl: hexan-3-one, 2-methylpentan-3-one, hexan-2-one, hexanal

4. Determine the identity of the molecule represented by the MS spectrum (graph)
   a.

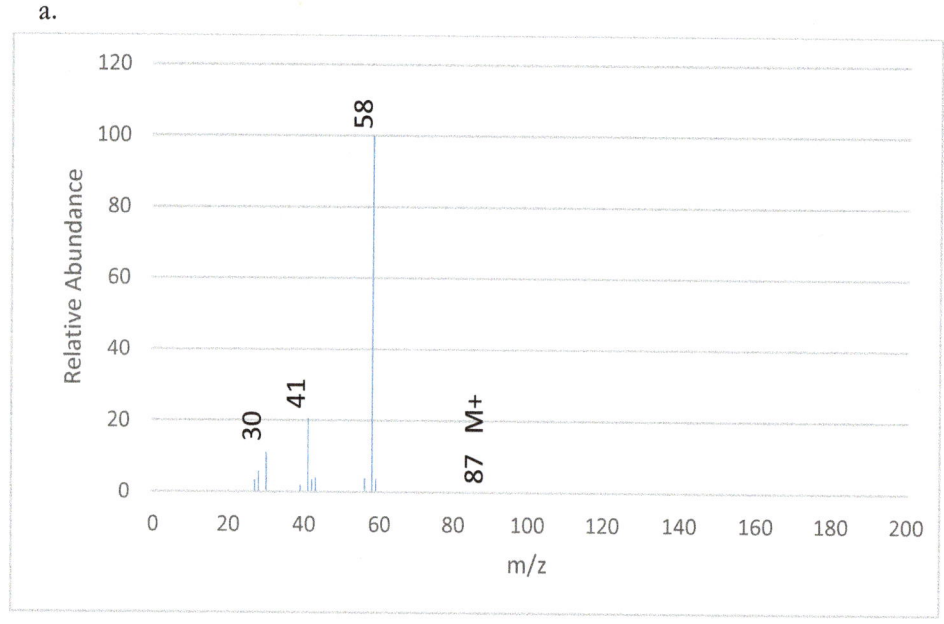

b.

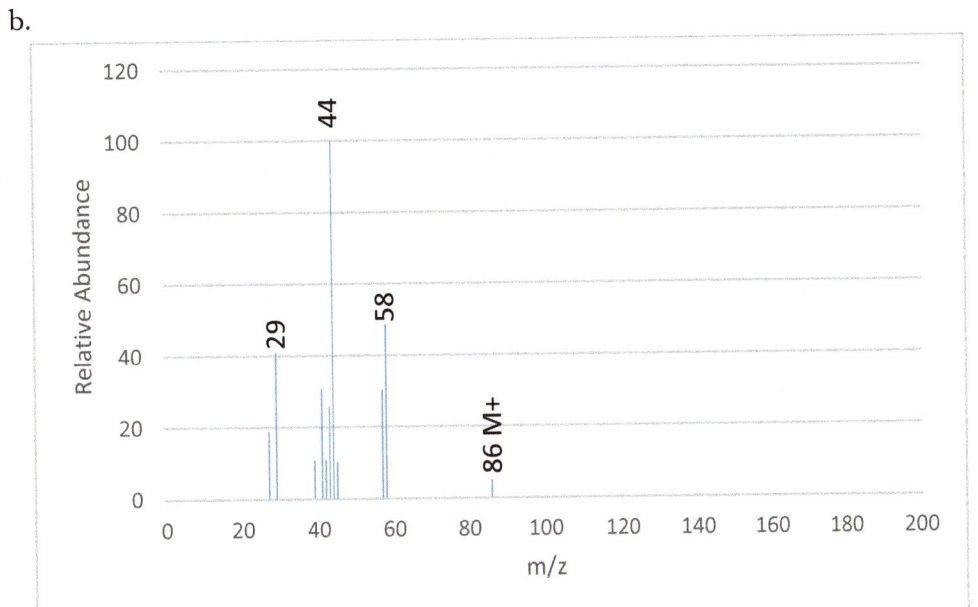

c.

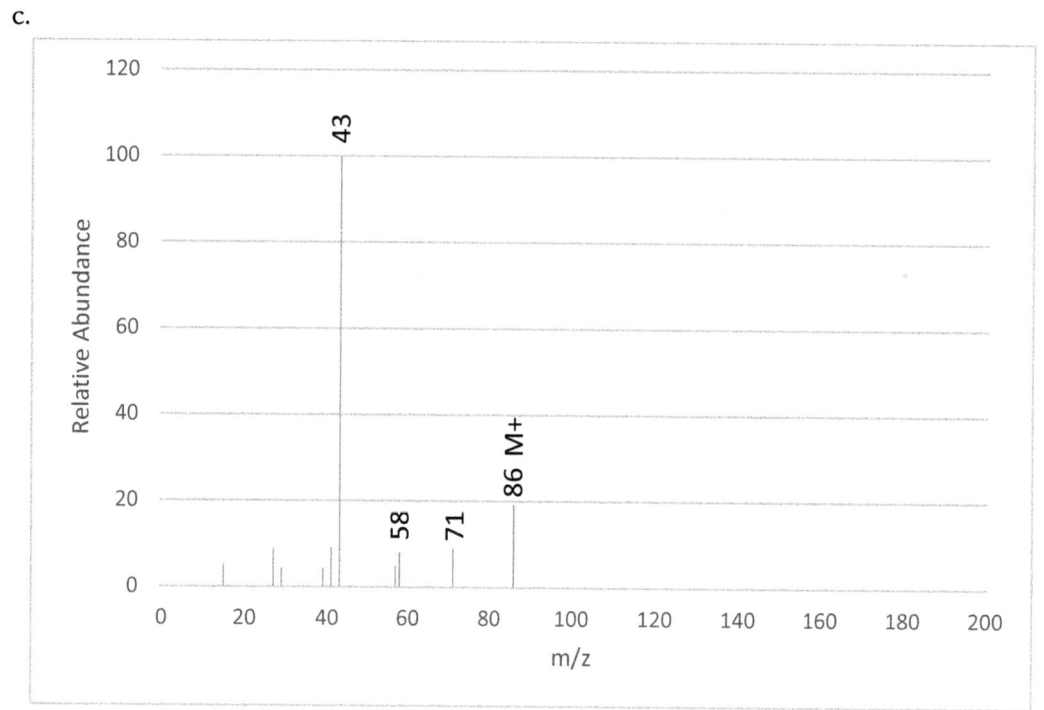

d. (ether)

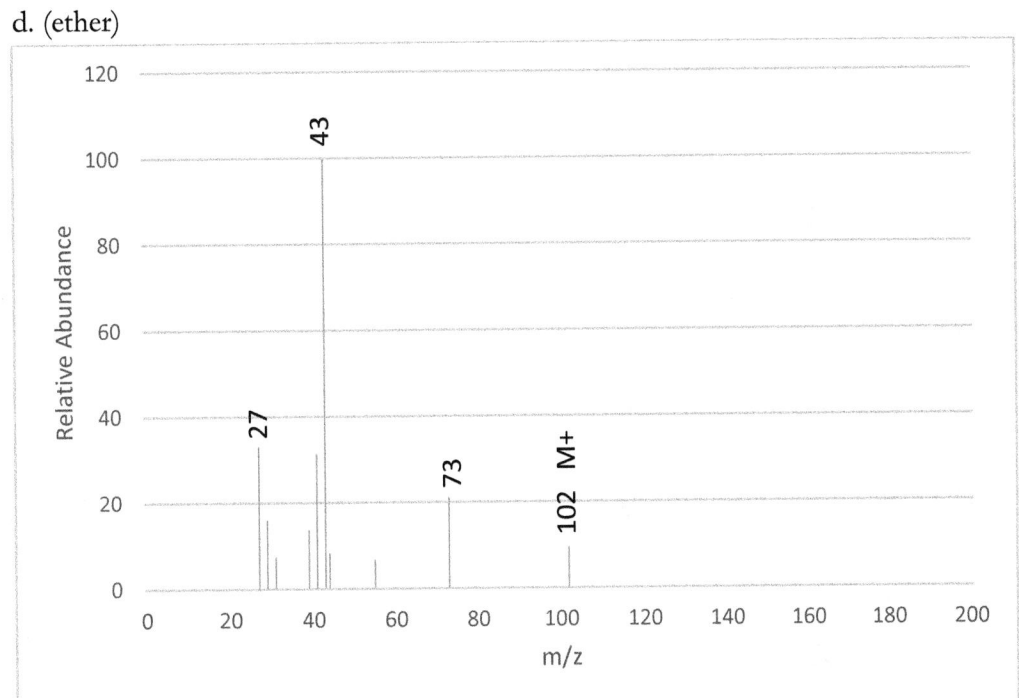

e.

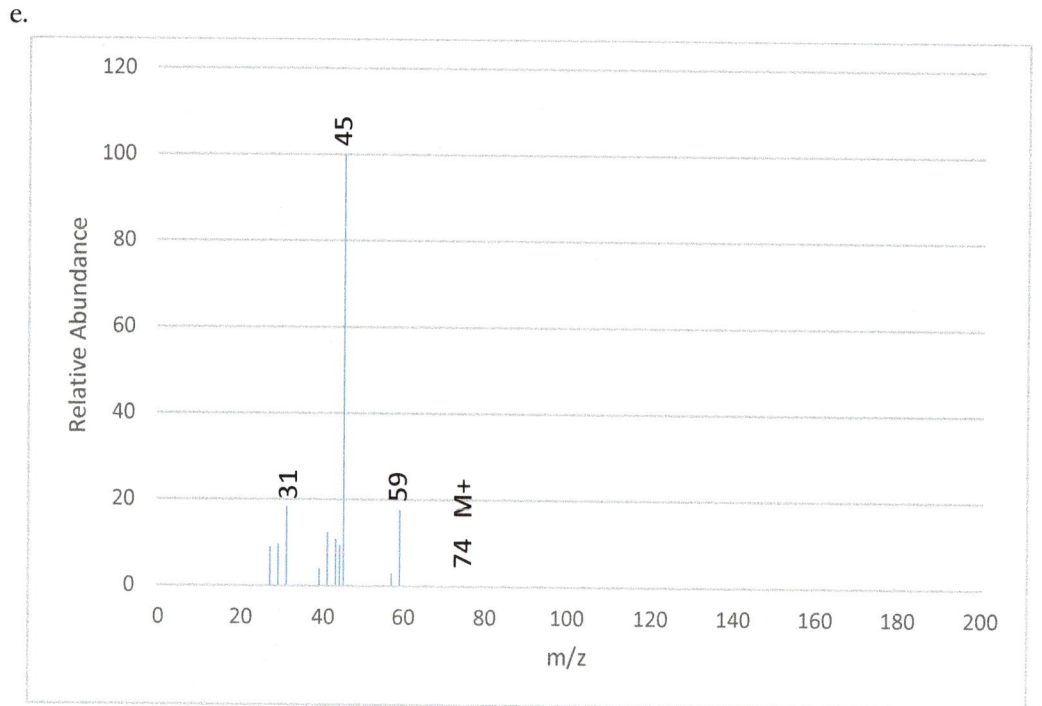

f.

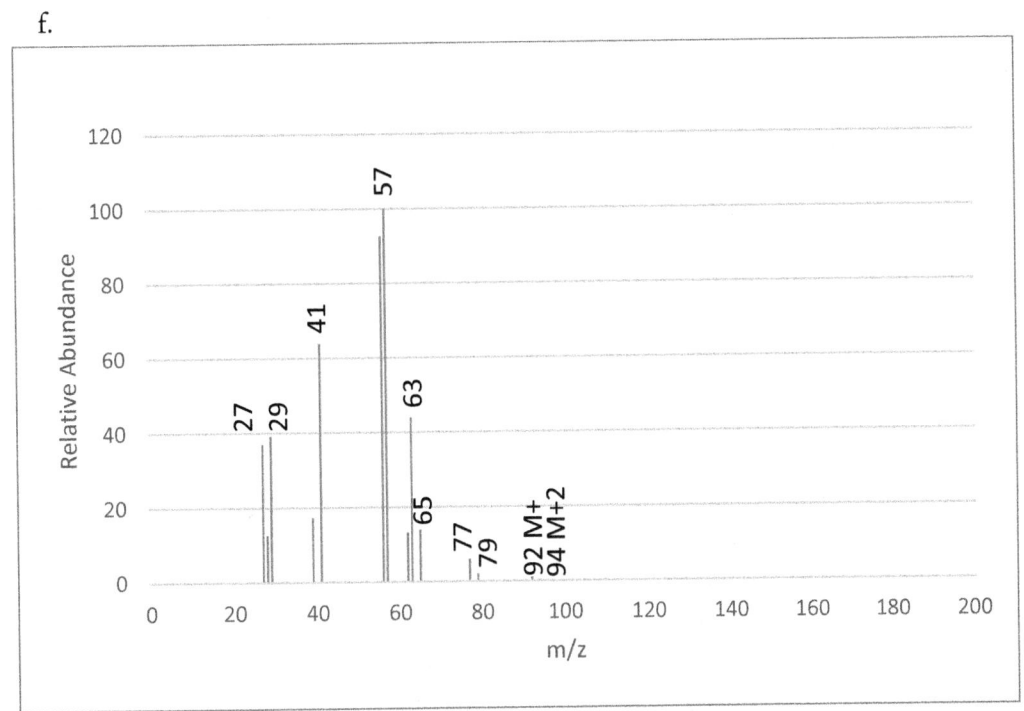

5. Determine the identity of each molecule based on the m/z values provided:

    a.   72, 85, 114 (M+)

       _____

    b.   69 (base peak), 148 (M+), 150 (same height as 148)

       _____

    c.   31, 59 (base peak), 73, 88 (M+)

       _____

    d.   72 (base peak), 86, 101 (M+)

       _____

e. 71, 77, 91, 93 (at 1/3 height of 91), 106 (M+), 108 (M+2 at 1/3 height of 106)

_____

f. alkane: 43, 57, 71 (base peak), 113, 127, 142 (M+)

_____

g. carbonyl: 29, 58 (base peak and M+)

_____

Mash-up 1

1. (Chapters 1 and 2) Assign an IUPAC name to the following:
   Stereochemistry:

   CH₃   CH₃   CH₃
   │   │   │
   H──┼──Cl   H──┼──Cl   H──┼──Cl
   │   │   │
   H──┼──OCH₂CH₃   H──┼──OCH₂CH₃   H──┼──OCH₂CH₃
   │   │   │
   CH₃   CH₃   CH₃

   IUPAC Name: _____

2. (Chapter 1) Assign an IUPAC name to the following:

$$H_3C-CH_2-CH_2-N-CH_2-CH(CH_3)-CH_3$$

with a CH_3 branch on the CH and an $H_3C-C(CH_3)_2-CH_3$ (tert-butyl) group on the N.

   IUPAC Name: _____

3. (Chapter 11) Assign an IUPAC and common name to the following:

$$Cl-C(=O)-CH_2-CH_2-CH(CH_3)-CH_3$$

   IUPAC Name: _____

$$Cl-C(=O)-CH_2-CH_2-CH(CH_3)-CH_3$$

   Common Name: _____

4. (Chapter 11) Assign both an IUPAC and a common name to the following:

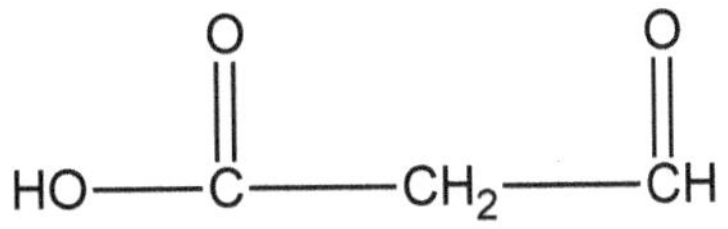

IUPAC Name: _____

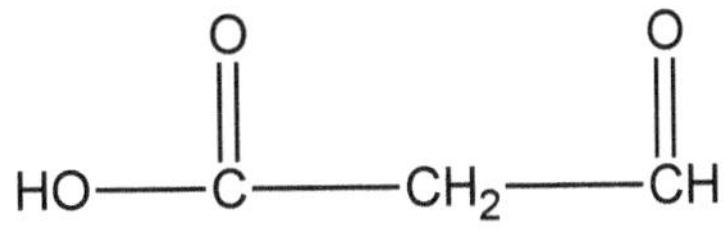

Common Name: _____

5. (Chapter 11) Assign a common name to the following:

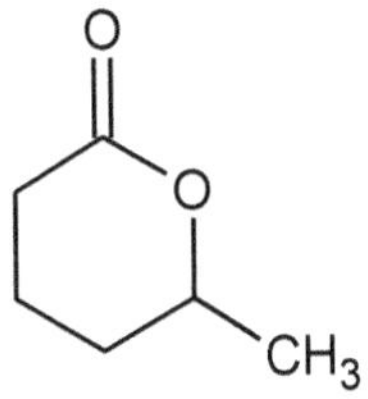

Common Name: _____

6. (Chapter 10) Assign a common name to the following:

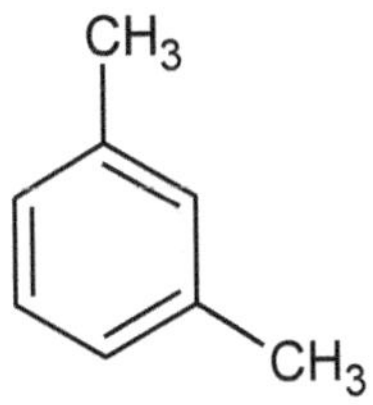

Common Name: _____

7. (Chapter 12) What two possibilities exist for reacting a carbonyl group that doesn't have a conjugated pi bond, why do these two possibilities exist, and under what conditions would you select each?

1.) _____

_____

2.) _____

_____

8. (Chapter 10) What type of instability will need to be addressed in the intermediate when an aromatic compound reacts with a very reactive electrophile?

_____

9. (Chapter 7) What type of instability will need to be addressed in the intermediate that is created when a pi bond with an electrophilic atom that has a non-bonded electron pair?

_____

10. (Chapter 13) What general resources could be used to bring electrons to a carbon that has a partial positive charge?

_____

_____

_____

11. (Chapter 9) Which is more stable and why?

$$CH_3$$
$$H_3C-CH-CH\cdot\cdot$$

$$CH_3$$
$$H_3C-C=CH\cdot\cdot$$

_____

12. (Chapters 8 and 10) Draw all resonance contributors for the following resonance-stabilized intermediate, label relative stabilities (major, minor, or contribute equally), and draw the structure of the actual molecule. Finally, state whether the molecule is aromatic, non-aromatic, or anti-aromatic.

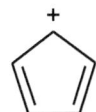

_____

(Chapters 5–13) For each of the following,

13. a. Label the reactive features, highlight the most reactive feature, and then highlight what it needs. Also, state if a carbocation, carbon radical, or carbanion will start to develop, and/or if aromatic character will be lost because of a reaction between these molecules. If a carbocation, carbon radical, or carbanion starts to develop, label where that will occur.

H_2SO_4

b. Use mechanism arrows to illustrate the reaction that occurs.

If applicable, use stabilization resources to deal with the carbocation, carbon radical, or carbanion that starts to develop during the reaction, and draw the structure of any resonance-stabilized intermediate.

Continue labelling and diagramming the reaction until you find the major stable product(s).

Finally, state the stereochemistry of the major product(s) and use either Fisher projection or perspective formula representations to illustrate that stereochemistry.

14. a. Label the reactive features, highlight the most reactive feature, and then highlight what it needs. Also, state if a carbocation, carbon radical, or carbanion will start to develop, and/or if aromatic character will be lost because of a reaction between these molecules. If a carbocation, carbon radical, or carbanion starts to develop, label where that will occur.

xs pentan-3-one + NaOH. When the reaction is complete, add acid.

b. Use mechanism arrows to illustrate the reaction that occurs.

If applicable, use stabilization resources to deal with the carbocation, carbon radical, or carbanion that starts to develop during the reaction, and draw the structure of any resonance-stabilized intermediate.

Continue labelling and diagramming the reaction until you find the major stable product(s).

Finally, state the stereochemistry of the major product(s) and use either Fisher projection or perspective formula representations to illustrate that stereochemistry.

15. a. Label the reactive features, highlight the most reactive feature, and then highlight what it needs. Also, state if a carbocation, carbon radical, or carbanion will start to develop, and/or if aromatic character will be lost because of a reaction between these molecules. If a carbocation, carbon radical, or carbanion starts to develop, label where that will occur.

pentan-3-one + $CH_3CH_2NH_2$ with a trace of acid

b. Use mechanism arrows to illustrate the reaction that occurs.

If applicable, use stabilization resources to deal with the carbocation, carbon radical, or carbanion that starts to develop during the reaction, and draw the structure of any resonance-stabilized intermediate.

Continue labelling and diagramming the reaction until you find the major stable product(s).

Finally, state the stereochemistry of the major product(s) and use either Fisher projection or perspective formula representations to illustrate that stereochemistry.

16. a. Label the reactive features, highlight the most reactive feature, and then highlight what it needs. Also, state if a carbocation, carbon radical, or carbanion will start to develop, and/or if aromatic character will be lost because of a reaction between these molecules. If a carbocation, carbon radical, or carbanion starts to develop, label where that will occur.

1,2-dimethylcyclopentene + BH_3 followed by H_2O_2, ^-OH, and H_2O

b. Use mechanism arrows to illustrate the reaction that occurs.

If applicable, use stabilization resources to deal with the carbocation, carbon radical, or carbanion that starts to develop during the reaction, and draw the structure of any resonance-stabilized intermediate.

Continue labelling and diagramming the reaction until you find the major stable product(s).

Finally, state the stereochemistry of the major product(s) and use either Fisher projection or perspective formula representations to illustrate that stereochemistry.

17. a. Label the reactive features, highlight the most reactive feature, and then highlight what it needs. Also, state if a carbocation, carbon radical, or carbanion will start to develop, and/or if aromatic character will be lost because of a reaction between these molecules. If a carbocation, carbon radical, or carbanion starts to develop, label where that will occur.

benzene + 2-methylpropanoyl chloride + AlCl₃

b. Use mechanism arrows to illustrate the reaction that occurs.

If applicable, use stabilization resources to deal with the carbocation, carbon radical, or carbanion that starts to develop during the reaction, and draw the structure of any resonance-stabilized intermediate.

Continue labelling and diagramming the reaction until you find the major stable product(s).

Finally, state the stereochemistry of the major product(s) and use either Fisher projection or perspective formula representations to illustrate that stereochemistry.

18. a. Label the reactive features, highlight the most reactive feature, and then highlight what it needs. Also, state if a carbocation, carbon radical, or carbanion will start to develop, and/or if aromatic character will be lost because of a reaction between these molecules. If a carbocation, carbon radical, or carbanion starts to develop, label where that will occur.

(R)-2-bromohexane with sodium methoxide in DMSO

b. Use mechanism arrows to illustrate the reaction that occurs.

If applicable, use stabilization resources to deal with the carbocation, carbon radical, or carbanion that starts to develop during the reaction, and draw the structure of any resonance-stabilized intermediate.

Continue labelling and diagramming the reaction until you find the major stable product(s).

Finally, state the stereochemistry of the major product(s) and use either Fisher projection or perspective formula representations to illustrate that stereochemistry.

19. a. Label the reactive features, highlight the most reactive feature, and then highlight what it needs. Also, state if a carbocation, carbon radical, or carbanion will start to develop, and/or if aromatic character will be lost because of a reaction between these molecules. If a carbocation, carbon radical, or carbanion starts to develop, label where that will occur.

(E)-4,5-dimethyloct-4-ene + Br_2 in ethanol

b. Use mechanism arrows to illustrate the reaction that occurs.

If applicable, use stabilization resources to deal with the carbocation, carbon radical, or carbanion that starts to develop during the reaction, and draw the structure of any resonance-stabilized intermediate.

Continue labelling and diagramming the reaction until you find the major stable product(s).

Finally, state the stereochemistry of the major product(s) and use either Fisher projection or perspective formula representations to illustrate that stereochemistry.

20. a. Label the reactive features, highlight the most reactive feature, and then highlight what it needs. Also, state if a carbocation, carbon radical, or carbanion will start to develop, and/or if aromatic character will be lost because of a reaction between these molecules. If a carbocation, carbon radical, or carbanion starts to develop, label where that will occur.

benzaldehyde + Cl_2 and $FeCl_3$

b. Use mechanism arrows to illustrate the reaction that occurs.

If applicable, use stabilization resources to deal with the carbocation, carbon radical, or carbanion that starts to develop during the reaction, and draw the structure of any resonance-stabilized intermediate.

Continue labelling and diagramming the reaction until you find the major stable product(s).

Finally, state the stereochemistry of the major product(s) and use either Fisher projection or perspective formula representations to illustrate that stereochemistry.

21. a. Label the reactive features, highlight the most reactive feature, and then highlight what it needs. Also, state if a carbocation, carbon radical, or carbanion will start to develop, and/or if aromatic character will be lost because of a reaction between these molecules. If a carbocation, carbon radical, or carbanion starts to develop, label where that will occur.

     m-bromoethylbenzene + Cl_2 and $FeCl_3$

     b. Use mechanism arrows to illustrate the reaction that occurs.

     If applicable, use stabilization resources to deal with the carbocation, carbon radical, or carbanion that starts to develop during the reaction, and draw the structure of any resonance-stabilized intermediate.

     Continue labelling and diagramming the reaction until you find the major stable product(s).

     Finally, state the stereochemistry of the major product(s) and use either Fisher projection or perspective formula representations to illustrate that stereochemistry.

Chapter 14

22. a. Label the reactive features, highlight the most reactive feature, and then highlight what it needs. Also, state if a carbocation, carbon radical, or carbanion will start to develop, and/or if aromatic character will be lost because of a reaction between these molecules. If a carbocation, carbon radical, or carbanion starts to develop, label where that will occur.

2-methylcyclopenta-1,3-diene with HBr

b. Use mechanism arrows to illustrate the reaction that occurs.

If applicable, use stabilization resources to deal with the carbocation, carbon radical, or carbanion that starts to develop during the reaction, and draw the structure of any resonance-stabilized intermediate.

Continue labelling and diagramming the reaction until you find the major stable product(s).

Finally, state the stereochemistry of the major product(s) and use either Fisher projection or perspective formula representations to illustrate that stereochemistry.

51

23. a. Label the reactive features, highlight the most reactive feature, and then highlight what it needs. Also, state if a carbocation, carbon radical, or carbanion will start to develop, and/or if aromatic character will be lost because of a reaction between these molecules. If a carbocation, carbon radical, or carbanion starts to develop, label where that will occur.

   hexan-1-ol with POCl₃ and xs pyridine

   b. Use mechanism arrows to illustrate the reaction that occurs.

   If applicable, use stabilization resources to deal with the carbocation, carbon radical, or carbanion that starts to develop during the reaction, and draw the structure of any resonance-stabilized intermediate.

   Continue labelling and diagramming the reaction until you find the major stable product(s).

   Finally, state the stereochemistry of the major product(s) and use either Fisher projection or perspective formula representations to illustrate that stereochemistry.

24. (Chapters 5–14) Identify the structure of the molecule represented by the MS spectrum shown below, assign both an IUPAC name and a derived name to that molecule, and then predict the aldol product resulting from the reaction of –OH with 2 equivalents of that compound:

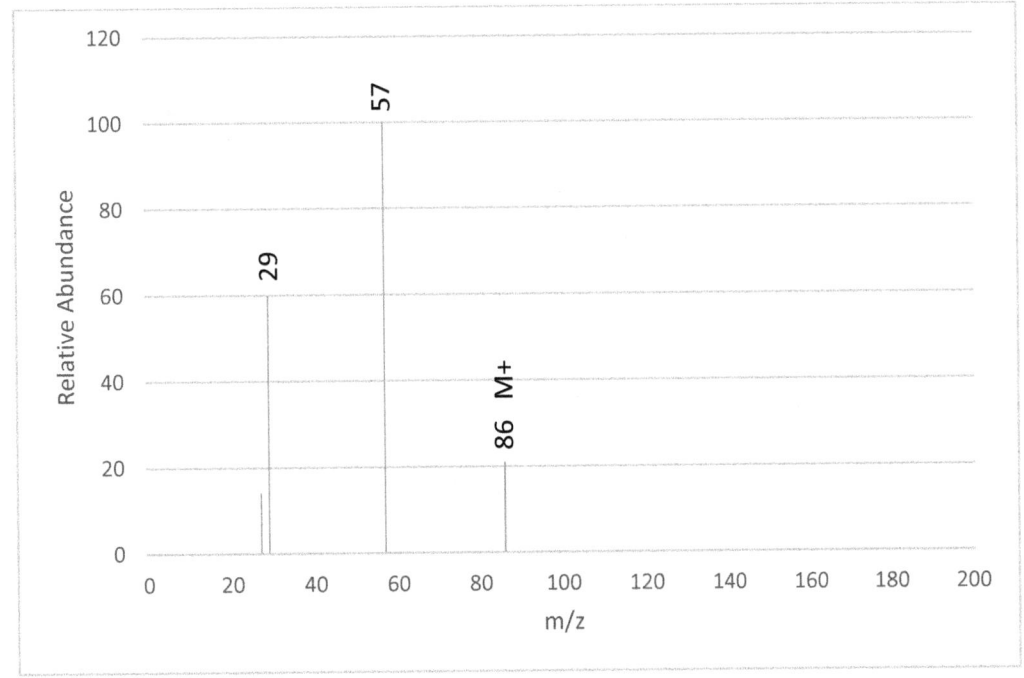

Structure: _____

IUPAC Name: _____

Derived Name: _____

Reaction:

Product: _____

Mash-up 2

1. (Chapters 1 and 2) Assign an IUPAC name to the following:

Stereochemistry:

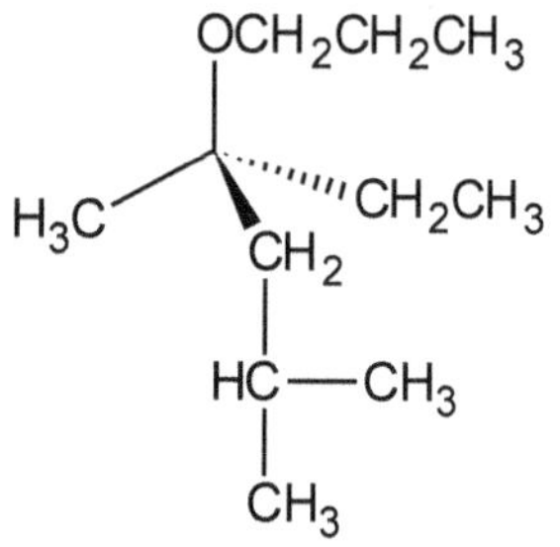

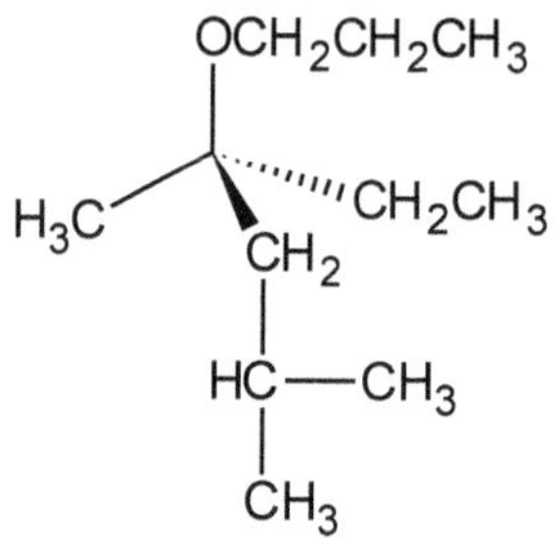

IUPAC Name: _____

2. (Chapters 1 and 3) Assign both an IUPAC name and a common name to the following:

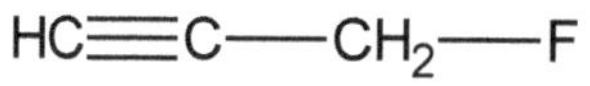

$$HC \equiv C - CH_2 - F$$

IUPAC Name: _____

$$HC \equiv C - CH_2 - F$$

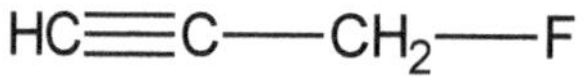

Common Name: _____

3. (Chapter 4) Assign both a mixed IUPAC and pure IUPAC name to the following:

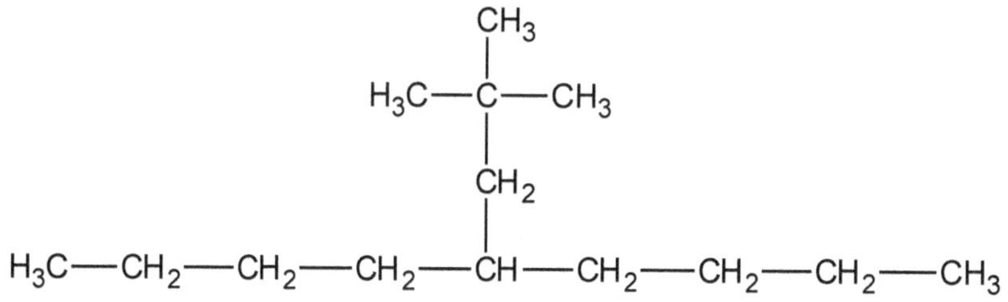

Mixed IUPAC Name: _____

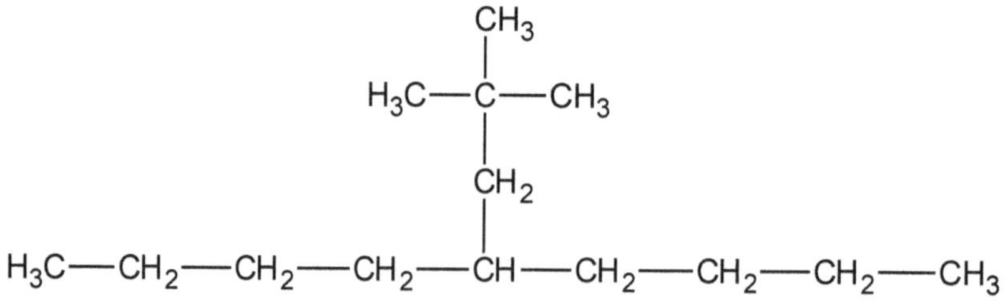

Pure IUPAC Name: _____

4. (Chapter 11) Assign both an IUPAC name and a derived name to the following:

$$H_3C-\overset{\overset{\displaystyle CH_3}{|}}{CH}-CH_2-\overset{\overset{\displaystyle O}{||}}{C}-\overset{\underset{\displaystyle CH_3}{|}}{CH}-CH_3$$

IUPAC Name: _____

$$H_3C-\overset{\overset{\displaystyle CH_3}{|}}{CH}-CH_2-\overset{\overset{\displaystyle O}{||}}{C}-\overset{\underset{\displaystyle CH_3}{|}}{CH}-CH_3$$

Derived Name: _____

5. (Chapter 11) Assign both an IUPAC name and a common name to the following:

$$\overset{\overset{\displaystyle O}{||}}{HC}-O-\overset{\overset{\displaystyle O}{||}}{CH}$$

IUPAC Name: _____

$$\overset{\overset{\displaystyle O}{||}}{HC}-O-\overset{\overset{\displaystyle O}{||}}{CH}$$

Common Name: _____

6. (Chapter 10) Assign a common name to the following:

$$HC=CH_2$$
Br

Common Name: _____

7. (Chapter 12) How can you tell when a reaction keeps going after a reaction involving a carbonyl carbon reaches an initial resolution? (In other words, what structure in the intermediate tells you that the reaction will continue until it forms either a Schiff base, acetal, or ketal?)

_____

_____

8. (Chapter 9) What type of instability will need to be addressed in the intermediate that is created when an enol is in acidic conditions?

_____

9. (Chapter 12) What type of instability will need to be addressed in the intermediate when a carbonyl carbon reacts directly with a nucleophile/base?

_____

10. (Chapters 6, 7, and 8) What general methods might potentially be used to stabilize a carbocation?

_____

_____

_____

_____

11. (Chapters 6 and 8) Which molecule in the following set is more stable and why?

HC—CH₃

H₂C—CH₃

_____

12. (Chapter 10) Is the following molecule aromatic, non-aromatic, or anti-aromatic?

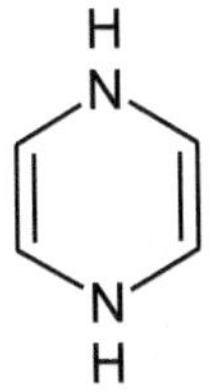

(Chapters 5–13) For each of the following,

13. a. Label the reactive features, highlight the most reactive feature, and then highlight what it needs. Also, state if a carbocation, carbon radical, or carbanion will start to develop, and/or if aromatic character will be lost because of a reaction between these molecules. If a carbocation, carbon radical, or carbanion starts to develop, label where that will occur.

1-methylcyclohexa-1,3-diene with HBr

b. Use mechanism arrows to illustrate the reaction that occurs.

If applicable, use stabilization resources to deal with the carbocation, carbon radical, or carbanion that starts to develop during the reaction, and draw the structure of any resonance-stabilized intermediate.

Continue labelling and diagramming the reaction until you find the major stable product(s).

Finally, state the stereochemistry of the major product(s) and use either Fisher projection or perspective formula representations to illustrate that stereochemistry.

14. a. Label the reactive features, highlight the most reactive feature, and then highlight what it needs. Also, state if a carbocation, carbon radical, or carbanion will start to develop, and/or if aromatic character will be lost because of a reaction between these molecules. If a carbocation, carbon radical, or carbanion starts to develop, label where that will occur.

benzenesulfonic acid + 2-chloro-3-methyl butane + $AlCl_3$

b. Use mechanism arrows to illustrate the reaction that occurs.

If applicable, use stabilization resources to deal with the carbocation, carbon radical, or carbanion that starts to develop during the reaction, and draw the structure of any resonance-stabilized intermediate.

Continue labelling and diagramming the reaction until you find the major stable product(s).

Finally, state the stereochemistry of the major product(s) and use either Fisher projection or perspective formula representations to illustrate that stereochemistry.

15. a. Label the reactive features, highlight the most reactive feature, and then highlight what it needs. Also, state if a carbocation, carbon radical, or carbanion will start to develop, and/or if aromatic character will be lost because of a reaction between these molecules. If a carbocation, carbon radical, or carbanion starts to develop, label where that will occur.

3-methylcyclohexene + $Hg(OAc)_2$ in methanol, followed by $NaBH_4$

b. Use mechanism arrows to illustrate the reaction that occurs.

If applicable, use stabilization resources to deal with the carbocation, carbon radical, or carbanion that starts to develop during the reaction, and draw the structure of any resonance-stabilized intermediate.

Continue labelling and diagramming the reaction until you find the major stable product(s).

Finally, state the stereochemistry of the major product(s) and use either Fisher projection or perspective formula representations to illustrate that stereochemistry.

16. a. Label the reactive features, highlight the most reactive feature, and then highlight what it needs. Also, state if a carbocation, carbon radical, or carbanion will start to develop, and/or if aromatic character will be lost because of a reaction between these molecules. If a carbocation, carbon radical, or carbanion starts to develop, label where that will occur.

(R)-3-methylpent-1-ene + Cl_2 in CH_2Cl_2

b. Use mechanism arrows to illustrate the reaction that occurs.

If applicable, use stabilization resources to deal with the carbocation, carbon radical, or carbanion that starts to develop during the reaction, and draw the structure of any resonance-stabilized intermediate.

Continue labelling and diagramming the reaction until you find the major stable product(s).

Finally, state the stereochemistry of the major product(s) and use either Fisher projection or perspective formula representations to illustrate that stereochemistry.

17. a. Label the reactive features, highlight the most reactive feature, and then highlight what it needs. Also, state if a carbocation, carbon radical, or carbanion will start to develop, and/or if aromatic character will be lost because of a reaction between these molecules. If a carbocation, carbon radical, or carbanion starts to develop, label where that will occur.

Cyclohexanone with 2 equivalents of methoxide in methanol

b. Use mechanism arrows to illustrate the reaction that occurs.

If applicable, use stabilization resources to deal with the carbocation, carbon radical, or carbanion that starts to develop during the reaction, and draw the structure of any resonance-stabilized intermediate.

Continue labelling and diagramming the reaction until you find the major stable product(s).

Finally, state the stereochemistry of the major product(s) and use either Fisher projection or perspective formula representations to illustrate that stereochemistry.

18. a. Label the reactive features, highlight the most reactive feature, and then highlight what it needs. Also, state if a carbocation, carbon radical, or carbanion will start to develop, and/or if aromatic character will be lost because of a reaction between these molecules. If a carbocation, carbon radical, or carbanion starts to develop, label where that will occur.

(1R, 3R)-3-chlorocyclohexanamine with sodium ethoxide in DMSO

b. Use mechanism arrows to illustrate the reaction that occurs.

If applicable, use stabilization resources to deal with the carbocation, carbon radical, or carbanion that starts to develop during the reaction, and draw the structure of any resonance-stabilized intermediate.

Continue labelling and diagramming the reaction until you find the major stable product(s).

Finally, state the stereochemistry of the major product(s) and use either Fisher projection or perspective formula representations to illustrate that stereochemistry.

19. a. Label the reactive features, highlight the most reactive feature, and then highlight what it needs. Also, state if a carbocation, carbon radical, or carbanion will start to develop, and/or if aromatic character will be lost because of a reaction between these molecules. If a carbocation, carbon radical, or carbanion starts to develop, label where that will occur.

       o-bromonitrobenzene with sodium methoxide

      b. Use mechanism arrows to illustrate the reaction that occurs.

      If applicable, use stabilization resources to deal with the carbocation, carbon radical, or carbanion that starts to develop during the reaction, and draw the structure of any resonance-stabilized intermediate.

      Continue labelling and diagramming the reaction until you find the major stable product(s).

      Finally, state the stereochemistry of the major product(s) and use either Fisher projection or perspective formula representations to illustrate that stereochemistry.

20. a. Label the reactive features, highlight the most reactive feature, and then highlight what it needs. Also, state if a carbocation, carbon radical, or carbanion will start to develop, and/or if aromatic character will be lost because of a reaction between these molecules. If a carbocation, carbon radical, or carbanion starts to develop, label where that will occur.

    phenol + 2-chloro-3-methyl butane + AlCl₃

    b. Use mechanism arrows to illustrate the reaction that occurs.

    If applicable, use stabilization resources to deal with the carbocation, carbon radical, or carbanion that starts to develop during the reaction, and draw the structure of any resonance-stabilized intermediate.

    Continue labelling and diagramming the reaction until you find the major stable product(s).

    Finally, state the stereochemistry of the major product(s) and use either Fisher projection or perspective formula representations to illustrate that stereochemistry.

21. a. Label the reactive features, highlight the most reactive feature, and then highlight what it needs. Also, state if a carbocation, carbon radical, or carbanion will start to develop, and/or if aromatic character will be lost because of a reaction between these molecules. If a carbocation, carbon radical, or carbanion starts to develop, label where that will occur.

    1-methylcyclohexa-1,3-diene with HBr

    b. Use mechanism arrows to illustrate the reaction that occurs.

    If applicable, use stabilization resources to deal with the carbocation, carbon radical, or carbanion that starts to develop during the reaction, and draw the structure of any resonance-stabilized intermediate.

    Continue labelling and diagramming the reaction until you find the major stable product(s).

    Finally, state the stereochemistry of the major product(s) and use either Fisher projection or perspective formula representations to illustrate that stereochemistry.

22. a. Label the reactive features, highlight the most reactive feature, and then highlight what it needs. Also, state if a carbocation, carbon radical, or carbanion will start to develop, and/or if aromatic character will be lost because of a reaction between these molecules. If a carbocation, carbon radical, or carbanion starts to develop, label where that will occur.

pentanoic acid + PBr₃

b. Use mechanism arrows to illustrate the reaction that occurs.

If applicable, use stabilization resources to deal with the carbocation, carbon radical, or carbanion that starts to develop during the reaction, and draw the structure of any resonance-stabilized intermediate.

Continue labelling and diagramming the reaction until you find the major stable product(s).

Finally, state the stereochemistry of the major product(s) and use either Fisher projection or perspective formula representations to illustrate that stereochemistry.

23. a. Label the reactive features, highlight the most reactive feature, and then highlight what it needs. Also, state if a carbocation, carbon radical, or carbanion will start to develop, and/or if aromatic character will be lost because of a reaction between these molecules. If a carbocation, carbon radical, or carbanion starts to develop, label where that will occur.

  p-ethylphenol + HNO_3 with tr. H_2SO_4

  b. Use mechanism arrows to illustrate the reaction that occurs.

  If applicable, use stabilization resources to deal with the carbocation, carbon radical, or carbanion that starts to develop during the reaction, and draw the structure of any resonance-stabilized intermediate.

  Continue labelling and diagramming the reaction until you find the major stable product(s).

  Finally, state the stereochemistry of the major product(s) and use either Fisher projection or perspective formula representations to illustrate that stereochemistry.

24. (Chapters 5–14) Identify the molecule represented by the following MS data, assign an IUPAC name and a derived name to the molecule, and then synthesize it from any hydrocarbon and any other combinations of reactants.

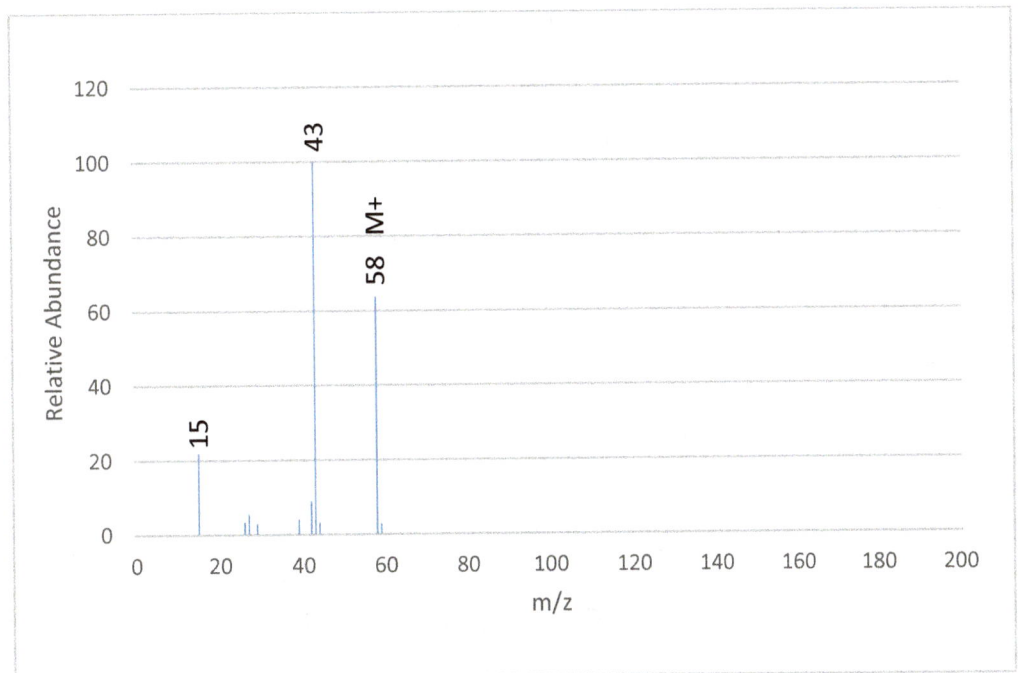

Structure: _____

IUPAC Name: _____

Derived Name: _____

Synthesis Strategy:

Mash-up 3

1. (Chapter 1) Assign an IUPAC name to the following:

$$H_3C-\overset{\displaystyle \overset{CH_3}{|}}{\underset{\displaystyle \underset{CH_2}{|}}{C}}-CH_3$$

$$H_3C-CH_2-CH_2-CH-CH_2-CH_2-CH_2-CH_3$$

IUPAC Name: _____

2. (Chapters 1 and 3) Assign both an IUPAC name and a common name to the following:

$$H_3C-C\equiv C-CH_2-CH_2-CH_3$$

IUPAC Name: _____

$$H_3C-C\equiv C-CH_2-CH_2-CH_3$$

Common Name: _____

3. (Chapter 11) Assign a common name to the following:

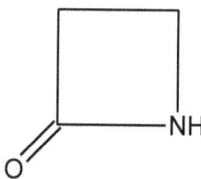

Common Name: _____

4. (Chapter 11) Assign an IUPAC name, a common name, and a derived name to the following:

$$H_3C-\overset{\displaystyle \overset{CH_3}{|}}{CH}-CH_2-CH_2-C\equiv N$$

IUPAC Name: _____

$$CH_3$$

$$H_3C—CH—CH_2—CH_2—C{\equiv}N$$

Common Name: _____

$$CH_3$$

$$H_3C—CH—CH_2—CH_2—C{\equiv}N$$

Derived Name: _____

5. (Chapter 10) Assign a common name to the following:

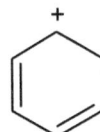

Common Name: _____

6. (Chapters 8 and 10) Draw all resonance contributors for the following resonance-stabilized intermediate, label relative stabilities (major, minor, or contribute equally), and draw the structure of the actual molecule. Finally, state whether the molecule is aromatic, non-aromatic, or anti-aromatic.

_____

7. (Chapter 12) What two possibilities exist for resolving a negative formal charge on oxygen in the intermediate created after a carbonyl carbon reacts with a nucleophile, and when would you use each?

1.) _____

_____

2.) _____

_____

8. (Chapter 13) What type of instability will need to be addressed in the intermediate that is created when an sp^3-hybridized carbon that has a leaving group is put into a protic polar solvent?

_____

9. (Chapter 6) Which molecule in the following set is more stable and why?

$$\underset{+}{H_3C-\overset{\overset{\displaystyle OH}{|}}{C}-CH_2-CH_3} \qquad \underset{+}{H_3C-\overset{\overset{\displaystyle NH_2}{|}}{C}-CH_2-CH_3}$$

_____

_____

10. (Chapter 9) What general methods could potentially be used to stabilize a carbanion?

_____

_____

_____

Chapter 14

(Chapters 5–13) For each of the following,

11. a. Label the reactive features, highlight the most reactive feature, and then highlight what it needs. Also, state if a carbocation, carbon radical, or carbanion will start to develop, and/or if aromatic character will be lost because of a reaction between these molecules. If a carbocation, carbon radical, or carbanion starts to develop, label where that will occur.

2-methylcyclopenta-1,3-diene + Br₂

b. Use mechanism arrows to illustrate the reaction that occurs.

If applicable, use stabilization resources to deal with the carbocation, carbon radical, or carbanion that starts to develop during the reaction, and draw the structure of any resonance-stabilized intermediate.

Continue labelling and diagramming the reaction until you find the major stable product(s).

Finally, state the stereochemistry of the major product(s) and use either Fisher projection or perspective formula representations to illustrate that stereochemistry.

12. a. Label the reactive features, highlight the most reactive feature, and then highlight what it needs. Also, state if a carbocation, carbon radical, or carbanion will start to develop, and/or if aromatic character will be lost because of a reaction between these molecules. If a carbocation, carbon radical, or carbanion starts to develop, label where that will occur.

CH3
|
N—CH3
[benzene ring]

H3C—CH—C—Cl   (CH3 above CH, O above C double bond)

AlCl3

H2C—CH3

b. Use mechanism arrows to illustrate the reaction that occurs.

If applicable, use stabilization resources to deal with the carbocation, carbon radical, or carbanion that starts to develop during the reaction, and draw the structure of any resonance-stabilized intermediate.

Continue labelling and diagramming the reaction until you find the major stable product(s).

Finally, state the stereochemistry of the major product(s) and use either Fisher projection or perspective formula representations to illustrate that stereochemistry.

13. a. Label the reactive features, highlight the most reactive feature, and then highlight what it needs. Also, state if a carbocation, carbon radical, or carbanion will start to develop, and/or if aromatic character will be lost because of a reaction between these molecules. If a carbocation, carbon radical, or carbanion starts to develop, label where that will occur.

ethyl pentanoate with sodium ethoxide in ethanol

b. Use mechanism arrows to illustrate the reaction that occurs.

If applicable, use stabilization resources to deal with the carbocation, carbon radical, or carbanion that starts to develop during the reaction, and draw the structure of any resonance-stabilized intermediate.

Continue labelling and diagramming the reaction until you find the major stable product(s).

Finally, state the stereochemistry of the major product(s) and use either Fisher projection or perspective formula representations to illustrate that stereochemistry.

14. a. Label the reactive features, highlight the most reactive feature, and then highlight what it needs. Also, state if a carbocation, carbon radical, or carbanion will start to develop, and/or if aromatic character will be lost because of a reaction between these molecules. If a carbocation, carbon radical, or carbanion starts to develop, label where that will occur.

(R)-pentan-2-ol with concentrated HCl

b. Use mechanism arrows to illustrate the reaction that occurs.

If applicable, use stabilization resources to deal with the carbocation, carbon radical, or carbanion that starts to develop during the reaction, and draw the structure of any resonance-stabilized intermediate.

Continue labelling and diagramming the reaction until you find the major stable product(s).

Finally, state the stereochemistry of the major product(s) and use either Fisher projection or perspective formula representations to illustrate that stereochemistry.

15. a. Label the reactive features, highlight the most reactive feature, and then highlight what it needs. Also, state if a carbocation, carbon radical, or carbanion will start to develop, and/or if aromatic character will be lost because of a reaction between these molecules. If a carbocation, carbon radical, or carbanion starts to develop, label where that will occur.

nitrobenzene + HNO_3 with tr. H_2SO_4

b. Use mechanism arrows to illustrate the reaction that occurs.

If applicable, use stabilization resources to deal with the carbocation, carbon radical, or carbanion that starts to develop during the reaction, and draw the structure of any resonance-stabilized intermediate.

Continue labelling and diagramming the reaction until you find the major stable product(s).

Finally, state the stereochemistry of the major product(s) and use either Fisher projection or perspective formula representations to illustrate that stereochemistry.

16. a. Label the reactive features, highlight the most reactive feature, and then highlight what it needs. Also, state if a carbocation, carbon radical, or carbanion will start to develop, and/or if aromatic character will be lost because of a reaction between these molecules. If a carbocation, carbon radical, or carbanion starts to develop, label where that will occur.

(2R,3R)-2-chloro-3-methylheptane in methanol

b. Use mechanism arrows to illustrate the reaction that occurs.

If applicable, use stabilization resources to deal with the carbocation, carbon radical, or carbanion that starts to develop during the reaction, and draw the structure of any resonance-stabilized intermediate.

Continue labelling and diagramming the reaction until you find the major stable product(s).

Finally, state the stereochemistry of the major product(s) and use either Fisher projection or perspective formula representations to illustrate that stereochemistry.

17. a. Label the reactive features, highlight the most reactive feature, and then highlight what it needs. Also, state if a carbocation, carbon radical, or carbanion will start to develop, and/or if aromatic character will be lost because of a reaction between these molecules. If a carbocation, carbon radical, or carbanion starts to develop, label where that will occur.

hex-3-yne + 1 equivalent of Br_2

b. Use mechanism arrows to illustrate the reaction that occurs.

If applicable, use stabilization resources to deal with the carbocation, carbon radical, or carbanion that starts to develop during the reaction, and draw the structure of any resonance-stabilized intermediate.

Continue labelling and diagramming the reaction until you find the major stable product(s).

Finally, state the stereochemistry of the major product(s) and use either Fisher projection or perspective formula representations to illustrate that stereochemistry.

18. a. Label the reactive features, highlight the most reactive feature, and then highlight what it needs. Also, state if a carbocation, carbon radical, or carbanion will start to develop, and/or if aromatic character will be lost because of a reaction between these molecules. If a carbocation, carbon radical, or carbanion starts to develop, label where that will occur.

    3,3-dimethylhept-1-ene + HBr and H_2O_2

    b. Use mechanism arrows to illustrate the reaction that occurs.

    If applicable, use stabilization resources to deal with the carbocation, carbon radical, or carbanion that starts to develop during the reaction, and draw the structure of any resonance-stabilized intermediate.

    Continue labelling and diagramming the reaction until you find the major stable product(s).

    Finally, state the stereochemistry of the major product(s) and use either Fisher projection or perspective formula representations to illustrate that stereochemistry.

19. a. Label the reactive features, highlight the most reactive feature, and then highlight what it needs. Also, state if a carbocation, carbon radical, or carbanion will start to develop, and/or if aromatic character will be lost because of a reaction between these molecules. If a carbocation, carbon radical, or carbanion starts to develop, label where that will occur.

HNO$_3$ with a trace of H$_2$SO$_4$

b. Use mechanism arrows to illustrate the reaction that occurs.

If applicable, use stabilization resources to deal with the carbocation, carbon radical, or carbanion that starts to develop during the reaction, and draw the structure of any resonance-stabilized intermediate.

Continue labelling and diagramming the reaction until you find the major stable product(s).

Finally, state the stereochemistry of the major product(s) and use either Fisher projection or perspective formula representations to illustrate that stereochemistry.

20. a. Label the reactive features, highlight the most reactive feature, and then highlight what it needs. Also, state if a carbocation, carbon radical, or carbanion will start to develop, and/or if aromatic character will be lost because of a reaction between these molecules. If a carbocation, carbon radical, or carbanion starts to develop, label where that will occur.

      ethyl isobutyl propyl amine with H_2O_2 in DMSO

      b. Use mechanism arrows to illustrate the reaction that occurs.

      If applicable, use stabilization resources to deal with the carbocation, carbon radical, or carbanion that starts to develop during the reaction, and draw the structure of any resonance-stabilized intermediate.

      Continue labelling and diagramming the reaction until you find the major stable product(s).

      Finally, state the stereochemistry of the major product(s) and use either Fisher projection or perspective formula representations to illustrate that stereochemistry.

Chapter 14

21. (Chapters 5–14) Identify the ether represented by the following MS data, assign a common name to the molecule, and then synthesize it from any hydrocarbon and any other combinations of reactants.

43 (base peak) 73, 87, 102 (M+)

Structure: _____

Common Name: _____

Synthesis Strategy:

Mash-up 4

1. (Chapters 1 and 2) Assign an IUPAC name to the following:

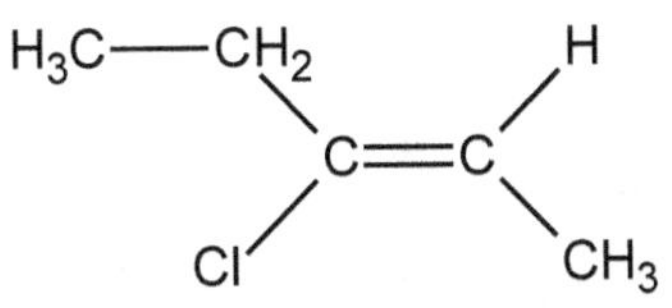

Stereochemistry:

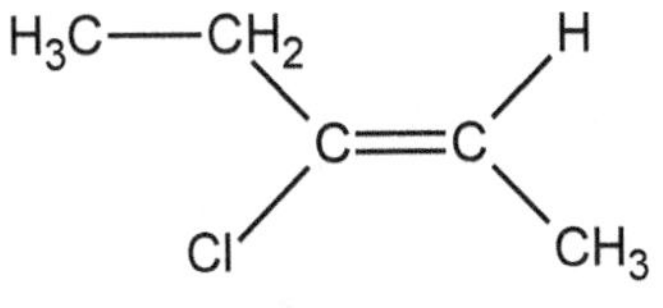

IUPAC Name: _____

2. (Chapters 1 and 3) Assign both an IUPAC name and a common name to the following:

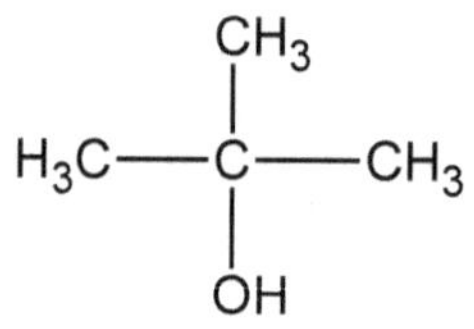

$$CH_3—C(CH_3)(OH)—CH_3$$

IUPAC Name: _____

$$CH_3—C(CH_3)(OH)—CH_3$$

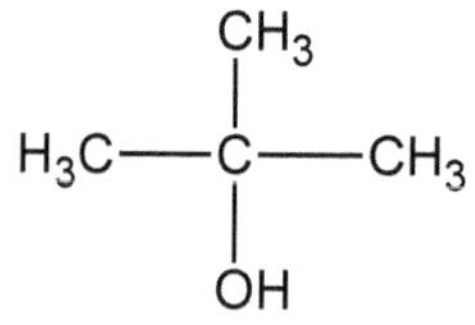

Common Name: _____

3. (Chapter 11) Assign an IUPAC name and a common name to the following:

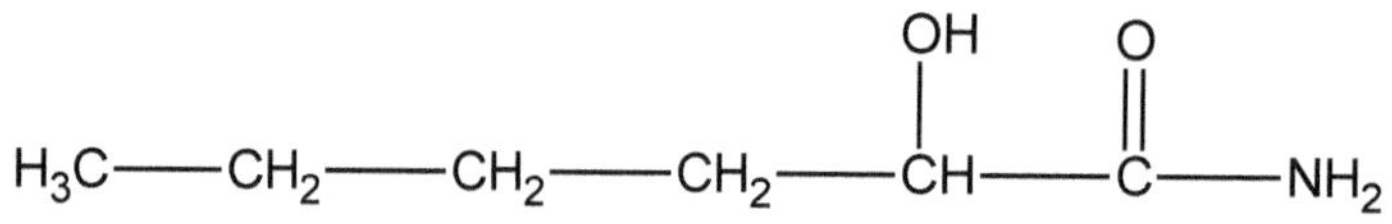

$$H_3C—CH_2—CH_2—CH_2—CH(OH)—C(=O)—NH_2$$

IUPAC Name: _____

$$H_3C—CH_2—CH_2—CH_2—CH(OH)—C(=O)—NH_2$$

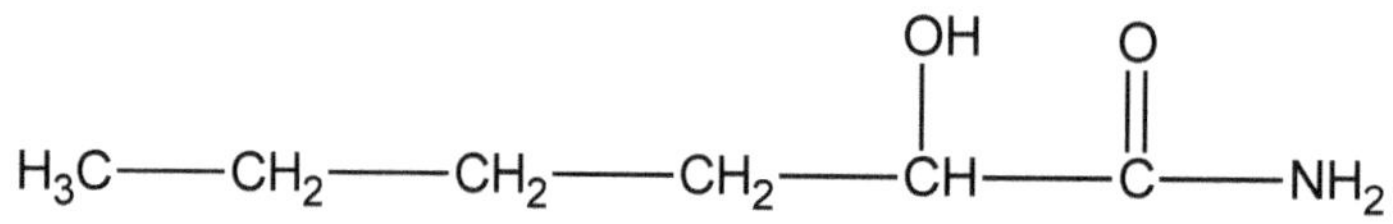

Common Name: _____

4. (Chapter 11) Assign an IUPAC name and a common name to the following:

$$HC(=O)—CH_2—CH(NH_2)—CH_3$$

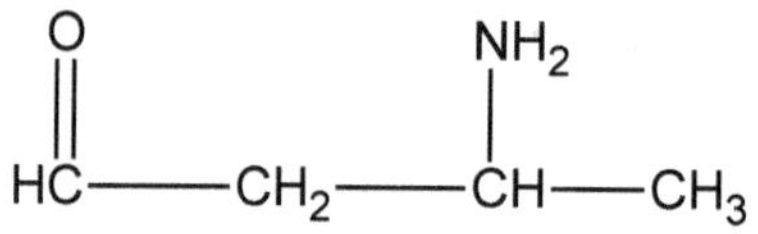

IUPAC Name: _____

O   NH$_2$

HC——CH$_2$——CH——CH$_3$

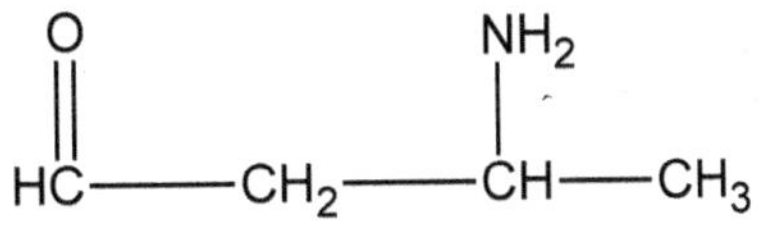

Common Name: _____

5. (Chapter 10) Assign a common name to the following:

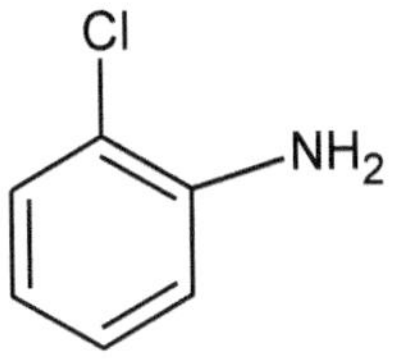

Cl

—NH$_2$

Common Name: _____

6. (Chapter 12) What two possibilities exist for reacting a carbonyl group when there is a conjugated pi bond, why do these two possibilities exist, and under what conditions would you select each?

1.) _____

_____

2.) _____

_____

7. (Chapter 6) What type of instability will need to be addressed in the intermediate that is created when a radical reacts with the bond electrons of a nucleophile?

_____

85

8. (Chapters 6 and 8) What general methods might potentially be used to stabilize a carbon radical?

_____

_____

_____

9. (Chapter 10) Is the following molecule aromatic, non-aromatic, or anti-aromatic?

_____

Chapter 14

(Chapters 5–13) For each of the following,

10. a. Label the reactive features, highlight the most reactive feature, and then highlight what it needs. Also, state if a carbocation, carbon radical, or carbanion will start to develop, and/or if aromatic character will be lost because of a reaction between these molecules. If a carbocation, carbon radical, or carbanion starts to develop, label where that will occur.

ethyl dipropyl amine + CH_3I followed by Ag_2O in H_2O

b. Use mechanism arrows to illustrate the reaction that occurs.

If applicable, use stabilization resources to deal with the carbocation, carbon radical, or carbanion that starts to develop during the reaction, and draw the structure of any resonance-stabilized intermediate.

Continue labelling and diagramming the reaction until you find the major stable product(s).

Finally, state the stereochemistry of the major product(s) and use either Fisher projection or perspective formula representations to illustrate that stereochemistry.

11. a. Label the reactive features, highlight the most reactive feature, and then highlight what it needs. Also, state if a carbocation, carbon radical, or carbanion will start to develop, and/or if aromatic character will be lost because of a reaction between these molecules. If a carbocation, carbon radical, or carbanion starts to develop, label where that will occur.

butanoyl chloride + 2 equivalents of ethyl amine

b. Use mechanism arrows to illustrate the reaction that occurs.

If applicable, use stabilization resources to deal with the carbocation, carbon radical, or carbanion that starts to develop during the reaction, and draw the structure of any resonance-stabilized intermediate.

Continue labelling and diagramming the reaction until you find the major stable product(s).

Finally, state the stereochemistry of the major product(s) and use either Fisher projection or perspective formula representations to illustrate that stereochemistry.

12. a. Label the reactive features, highlight the most reactive feature, and then highlight what it needs. Also, state if a carbocation, carbon radical, or carbanion will start to develop, and/or if aromatic character will be lost because of a reaction between these molecules. If a carbocation, carbon radical, or carbanion starts to develop, label where that will occur.

4-methylpenta-1,3-diene with $CH\equiv CCH=O$

b. Use mechanism arrows to illustrate the reaction that occurs.

If applicable, use stabilization resources to deal with the carbocation, carbon radical, or carbanion that starts to develop during the reaction, and draw the structure of any resonance-stabilized intermediate.

Continue labelling and diagramming the reaction until you find the major stable product(s).

Finally, state the stereochemistry of the major product(s) and use either Fisher projection or perspective formula representations to illustrate that stereochemistry.

13. a. Label the reactive features, highlight the most reactive feature, and then highlight what it needs. Also, state if a carbocation, carbon radical, or carbanion will start to develop, and/or if aromatic character will be lost because of a reaction between these molecules. If a carbocation, carbon radical, or carbanion starts to develop, label where that will occur.

m-propylbenzaldehyde + HNO_3 with tr. H_2SO_4

b. Use mechanism arrows to illustrate the reaction that occurs.

If applicable, use stabilization resources to deal with the carbocation, carbon radical, or carbanion that starts to develop during the reaction, and draw the structure of any resonance-stabilized intermediate.

Continue labelling and diagramming the reaction until you find the major stable product(s).

Finally, state the stereochemistry of the major product(s) and use either Fisher projection or perspective formula representations to illustrate that stereochemistry.

14. a. Label the reactive features, highlight the most reactive feature, and then highlight what it needs. Also, state if a carbocation, carbon radical, or carbanion will start to develop, and/or if aromatic character will be lost because of a reaction between these molecules. If a carbocation, carbon radical, or carbanion starts to develop, label where that will occur.

(Z)-3-methylhept-3-ene in ethanol with a tr. of acid

b. Use mechanism arrows to illustrate the reaction that occurs.

If applicable, use stabilization resources to deal with the carbocation, carbon radical, or carbanion that starts to develop during the reaction, and draw the structure of any resonance-stabilized intermediate.

Continue labelling and diagramming the reaction until you find the major stable product(s).

Finally, state the stereochemistry of the major product(s) and use either Fisher projection or perspective formula representations to illustrate that stereochemistry.

15. a. Label the reactive features, highlight the most reactive feature, and then highlight what it needs. Also, state if a carbocation, carbon radical, or carbanion will start to develop, and/or if aromatic character will be lost because of a reaction between these molecules. If a carbocation, carbon radical, or carbanion starts to develop, label where that will occur.

b. Use mechanism arrows to illustrate the reaction that occurs.

If applicable, use stabilization resources to deal with the carbocation, carbon radical, or carbanion that starts to develop during the reaction, and draw the structure of any resonance-stabilized intermediate.

Continue labelling and diagramming the reaction until you find the major stable product(s).

Finally, state the stereochemistry of the major product(s) and use either Fisher projection or perspective formula representations to illustrate that stereochemistry.

16. a. Label the reactive features, highlight the most reactive feature, and then highlight what it needs. Also, state if a carbocation, carbon radical, or carbanion will start to develop, and/or if aromatic character will be lost because of a reaction between these molecules. If a carbocation, carbon radical, or carbanion starts to develop, label where that will occur.

3,3-dimethylcyclohexene + Br₂ in ethanol

b. Use mechanism arrows to illustrate the reaction that occurs.

If applicable, use stabilization resources to deal with the carbocation, carbon radical, or carbanion that starts to develop during the reaction, and draw the structure of any resonance-stabilized intermediate.

Continue labelling and diagramming the reaction until you find the major stable product(s).

Finally, state the stereochemistry of the major product(s) and use either Fisher projection or perspective formula representations to illustrate that stereochemistry.

17. a. Label the reactive features, highlight the most reactive feature, and then highlight what it needs. Also, state if a carbocation, carbon radical, or carbanion will start to develop, and/or if aromatic character will be lost because of a reaction between these molecules. If a carbocation, carbon radical, or carbanion starts to develop, label where that will occur.

       tert-butyl alcohol with concentrated H_3PO_4

b. Use mechanism arrows to illustrate the reaction that occurs.

If applicable, use stabilization resources to deal with the carbocation, carbon radical, or carbanion that starts to develop during the reaction, and draw the structure of any resonance-stabilized intermediate.

Continue labelling and diagramming the reaction until you find the major stable product(s).

Finally, state the stereochemistry of the major product(s) and use either Fisher projection or perspective formula representations to illustrate that stereochemistry.

18. a. Label the reactive features, highlight the most reactive feature, and then highlight what it needs. Also, state if a carbocation, carbon radical, or carbanion will start to develop, and/or if aromatic character will be lost because of a reaction between these molecules. If a carbocation, carbon radical, or carbanion starts to develop, label where that will occur.

1-chloropentane + potassium hydroxide in DMSO

b. Use mechanism arrows to illustrate the reaction that occurs.

If applicable, use stabilization resources to deal with the carbocation, carbon radical, or carbanion that starts to develop during the reaction, and draw the structure of any resonance-stabilized intermediate.

Continue labelling and diagramming the reaction until you find the major stable product(s).

Finally, state the stereochemistry of the major product(s) and use either Fisher projection or perspective formula representations to illustrate that stereochemistry.

19. a. Label the reactive features, highlight the most reactive feature, and then highlight what it needs. Also, state if a carbocation, carbon radical, or carbanion will start to develop, and/or if aromatic character will be lost because of a reaction between these molecules. If a carbocation, carbon radical, or carbanion starts to develop, label where that will occur.

$$HN-CH_3$$

$$H_2SO_4$$

$$CH_2-CH_3$$

b. Use mechanism arrows to illustrate the reaction that occurs.

If applicable, use stabilization resources to deal with the carbocation, carbon radical, or carbanion that starts to develop during the reaction, and draw the structure of any resonance-stabilized intermediate.

Continue labelling and diagramming the reaction until you find the major stable product(s).

Finally, state the stereochemistry of the major product(s) and use either Fisher projection or perspective formula representations to illustrate that stereochemistry.

20. (Chapters 5–14) Identify the molecule represented by the following MS data, assign an IUPAC name and a common name to the molecule, and then synthesize it from any hydrocarbon and any other combinations of reactants.

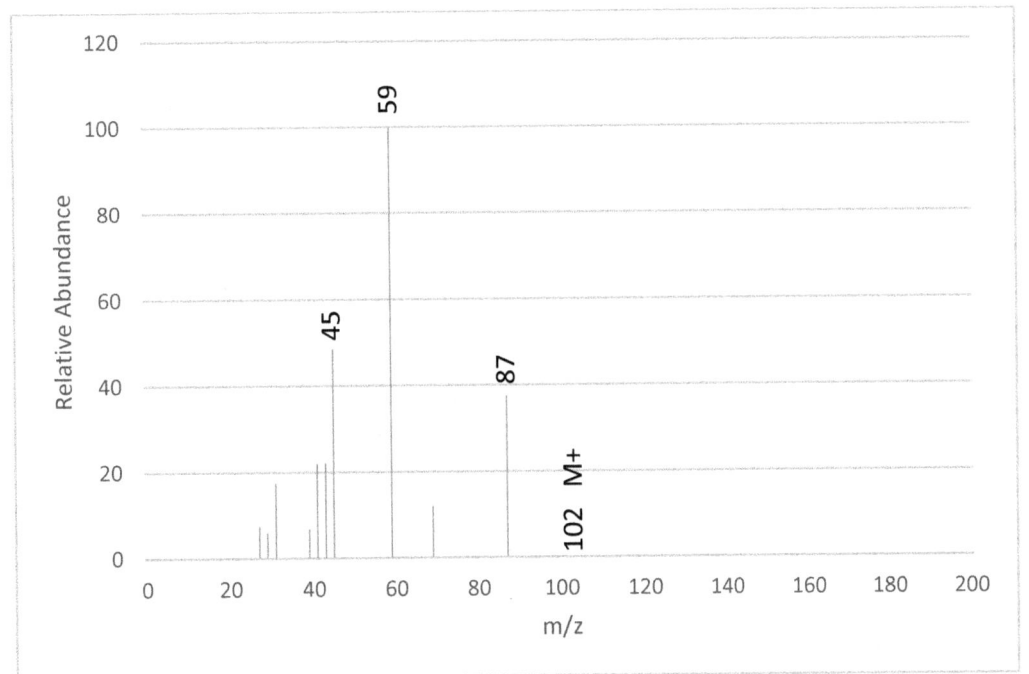

Structure: _____

IUPAC Name: _____

Common Name: _____

Synthesis Strategy:

Mash-up 5

1. (Chapters 1 and 3) Assign both an IUPAC name and a common name to the following:

$$H_3C-CH_2-O-\underset{\underset{CH_3}{|}}{\overset{\overset{CH_3}{|}}{C}}-CH_3$$

IUPAC Name: _____

$$H_3C-CH_2-O-\underset{\underset{CH_3}{|}}{\overset{\overset{CH_3}{|}}{C}}-CH_3$$

Common Name: _____

2. (Chapters 1 and 3) Assign both an IUPAC name and a common name to the following:

$$H_3C-\underset{\underset{CH_3}{|}}{CH}-Cl$$

IUPAC Name: _____

$$H_3C-\underset{\underset{CH_3}{|}}{CH}-Cl$$

Common Name: _____

3. (Chapter 11) Assign both an IUPAC name and a common name to the following:

$$H_3C-\underset{\underset{OH}{|}}{CH}-CH_2-CH_2-\underset{\underset{Cl}{|}}{CH}-\overset{\overset{O}{\|}}{C}-O-CH_3$$

IUPAC Name: _____

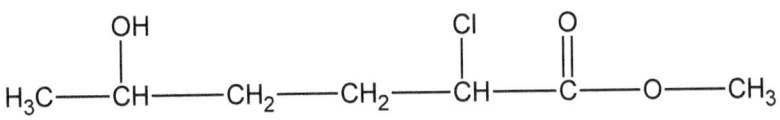

Common Name: _____

4. (Chapter 10) Assign a common name to the following:

Common Name: _____

5. (Chapter 6) What type of instability will need to be addressed in the intermediate that is created when a pi bond reacts with a hydrogen that has a partial positive charge?

_____

6. (Chapter 6) Which molecule in the following set is more stable and why?

$$H_3C—\overset{\overset{\displaystyle CH_3}{|}}{CH}—CH_2—\overset{\displaystyle \bullet}{CH_2} \qquad H_3C—\overset{\overset{\displaystyle CH_3}{|}}{CH}—\overset{\displaystyle \bullet}{CH}—CH_3$$

_____

22. Which of the following is the stronger acid, and why?

_____

_____

(Chapters 5–13) For each of the following,

7. a. Label the reactive features, highlight the most reactive feature, and then highlight what it needs. Also, state if a carbocation, carbon radical, or carbanion will start to develop, and/or if aromatic character will be lost because of a reaction between these molecules. If a carbocation, carbon radical, or carbanion starts to develop, label where that will occur.

        (R)-pentan-2-ol with TsCl and NaSCH₃

       b. Use mechanism arrows to illustrate the reaction that occurs.

       If applicable, use stabilization resources to deal with the carbocation, carbon radical, or carbanion that starts to develop during the reaction, and draw the structure of any resonance-stabilized intermediate.

       Continue labelling and diagramming the reaction until you find the major stable product(s).

       Finally, state the stereochemistry of the major product(s) and use either Fisher projection or perspective formula representations to illustrate that stereochemistry.

8. a. Label the reactive features, highlight the most reactive feature, and then highlight what it needs. Also, state if a carbocation, carbon radical, or carbanion will start to develop, and/or if aromatic character will be lost because of a reaction between these molecules. If a carbocation, carbon radical, or carbanion starts to develop, label where that will occur.

bromobenzene + H_2SO_4

b. Use mechanism arrows to illustrate the reaction that occurs.

If applicable, use stabilization resources to deal with the carbocation, carbon radical, or carbanion that starts to develop during the reaction, and draw the structure of any resonance-stabilized intermediate.

Continue labelling and diagramming the reaction until you find the major stable product(s).

Finally, state the stereochemistry of the major product(s) and use either Fisher projection or perspective formula representations to illustrate that stereochemistry.

9. a. Label the reactive features, highlight the most reactive feature, and then highlight what it needs. Also, state if a carbocation, carbon radical, or carbanion will start to develop, and/or if aromatic character will be lost because of a reaction between these molecules. If a carbocation, carbon radical, or carbanion starts to develop, label where that will occur.

       3-bromo-3-methylpentane with potassium methoxide

      b. Use mechanism arrows to illustrate the reaction that occurs.

      If applicable, use stabilization resources to deal with the carbocation, carbon radical, or carbanion that starts to develop during the reaction, and draw the structure of any resonance-stabilized intermediate.

      Continue labelling and diagramming the reaction until you find the major stable product(s).

      Finally, state the stereochemistry of the major product(s) and use either Fisher projection or perspective formula representations to illustrate that stereochemistry.

10. a. Label the reactive features, highlight the most reactive feature, and then highlight what it needs. Also, state if a carbocation, carbon radical, or carbanion will start to develop, and/or if aromatic character will be lost because of a reaction between these molecules. If a carbocation, carbon radical, or carbanion starts to develop, label where that will occur.

penta-1,3-diene with Br₂

b. Use mechanism arrows to illustrate the reaction that occurs.

If applicable, use stabilization resources to deal with the carbocation, carbon radical, or carbanion that starts to develop during the reaction, and draw the structure of any resonance-stabilized intermediate.

Continue labelling and diagramming the reaction until you find the major stable product(s).

Finally, state the stereochemistry of the major product(s) and use either Fisher projection or perspective formula representations to illustrate that stereochemistry.

11. a. Label the reactive features, highlight the most reactive feature, and then highlight what it needs. Also, state if a carbocation, carbon radical, or carbanion will start to develop, and/or if aromatic character will be lost because of a reaction between these molecules. If a carbocation, carbon radical, or carbanion starts to develop, label where that will occur.

2-methylcyclopenta-1,3-diene with $CH_3CH=CHC\equiv N$

b. Use mechanism arrows to illustrate the reaction that occurs.

If applicable, use stabilization resources to deal with the carbocation, carbon radical, or carbanion that starts to develop during the reaction, and draw the structure of any resonance-stabilized intermediate.

Continue labelling and diagramming the reaction until you find the major stable product(s).

Finally, state the stereochemistry of the major product(s) and use either Fisher projection or perspective formula representations to illustrate that stereochemistry.

12. a. Label the reactive features, highlight the most reactive feature, and then highlight what it needs. Also, state if a carbocation, carbon radical, or carbanion will start to develop, and/or if aromatic character will be lost because of a reaction between these molecules. If a carbocation, carbon radical, or carbanion starts to develop, label where that will occur.

1-methylcyclopentene + HBr

b. Use mechanism arrows to illustrate the reaction that occurs.

If applicable, use stabilization resources to deal with the carbocation, carbon radical, or carbanion that starts to develop during the reaction, and draw the structure of any resonance-stabilized intermediate.

Continue labelling and diagramming the reaction until you find the major stable product(s).

Finally, state the stereochemistry of the major product(s) and use either Fisher projection or perspective formula representations to illustrate that stereochemistry.

13. a. Label the reactive features, highlight the most reactive feature, and then highlight what it needs. Also, state if a carbocation, carbon radical, or carbanion will start to develop, and/or if aromatic character will be lost because of a reaction between these molecules. If a carbocation, carbon radical, or carbanion starts to develop, label where that will occur.

        p-ethylaniline + Cl_2 and $FeCl_3$

       b. Use mechanism arrows to illustrate the reaction that occurs.

       If applicable, use stabilization resources to deal with the carbocation, carbon radical, or carbanion that starts to develop during the reaction, and draw the structure of any resonance-stabilized intermediate.

       Continue labelling and diagramming the reaction until you find the major stable product(s).

       Finally, state the stereochemistry of the major product(s) and use either Fisher projection or perspective formula representations to illustrate that stereochemistry.

14. a. Label the reactive features, highlight the most reactive feature, and then highlight what it needs. Also, state if a carbocation, carbon radical, or carbanion will start to develop, and/or if aromatic character will be lost because of a reaction between these molecules. If a carbocation, carbon radical, or carbanion starts to develop, label where that will occur.

hex-3-en-2-one with sodium hydroxide

b. Use mechanism arrows to illustrate the reaction that occurs.

If applicable, use stabilization resources to deal with the carbocation, carbon radical, or carbanion that starts to develop during the reaction, and draw the structure of any resonance-stabilized intermediate.

Continue labelling and diagramming the reaction until you find the major stable product(s).

Finally, state the stereochemistry of the major product(s) and use either Fisher projection or perspective formula representations to illustrate that stereochemistry.

15. a. Label the reactive features, highlight the most reactive feature, and then highlight what it needs. Also, state if a carbocation, carbon radical, or carbanion will start to develop, and/or if aromatic character will be lost because of a reaction between these molecules. If a carbocation, carbon radical, or carbanion starts to develop, label where that will occur.

Cl_2 and $FeCl_3$

    b. Use mechanism arrows to illustrate the reaction that occurs.

    If applicable, use stabilization resources to deal with the carbocation, carbon radical, or carbanion that starts to develop during the reaction, and draw the structure of any resonance-stabilized intermediate.

    Continue labelling and diagramming the reaction until you find the major stable product(s).

    Finally, state the stereochemistry of the major product(s) and use either Fisher projection or perspective formula representations to illustrate that stereochemistry.

21. (Chapters 5–14) Identify the molecule represented by the following MS data, assign a common name to the molecule, and then synthesize it from any hydrocarbon and any other combinations of reactants.

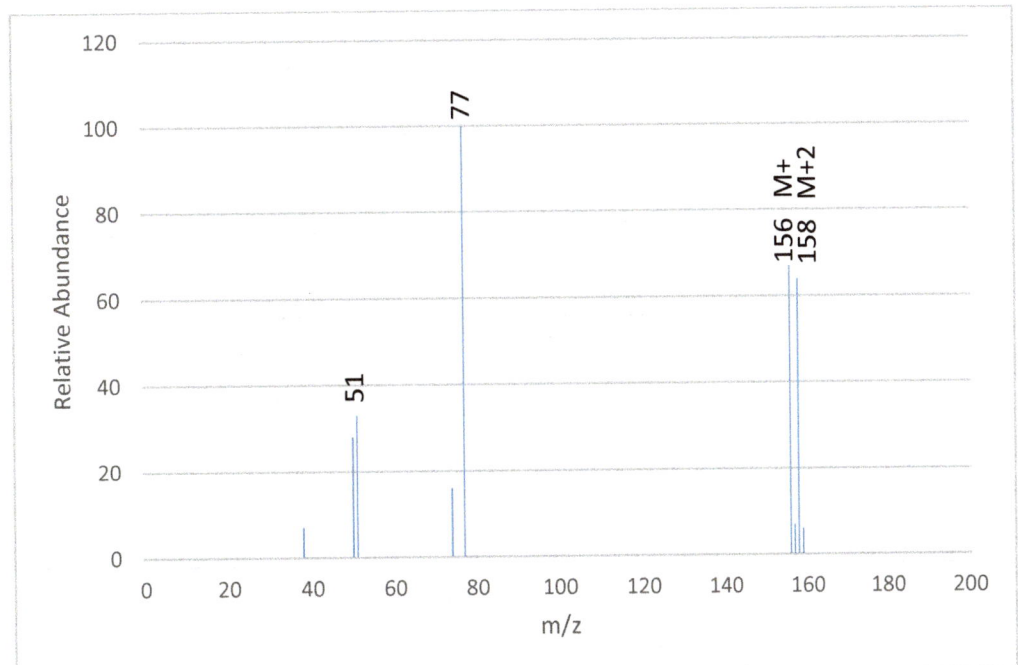

Structure: _____

Common Name: _____

Synthesis Strategy:

Summary of Concepts and Analysis Methods

Summary of the Chapter:

_____

_____

_____

_____

_____

_____

_____

Questions You Should Ask In Class:

_____

_____

_____

_____

_____

Chapter 14

Common Mistakes You Tend to Make but Want to Avoid in the Future:

_____

_____

_____

_____

_____

Types of Problems Needed for Targeted Practice:

_____

_____

_____

_____

_____

_____

Chapter 15

Understanding How to Analyze Structures of Products (Part 2): Interpreting Infrared Spectroscopy (IR), Polarimetry, and Ultraviolet-Visible Spectrophotometry (UV-Vis)

Key Concepts

IR:

Infrared spectroscopy works on the principle that electromagnetic radiation is absorbed when the frequency of bond vibration matches the frequency of the radiation. This match occurs in the IR portion of the electromagnetic spectrum.

The shorter the bond, the faster it vibrates, and therefore, the higher the frequency of radiation it absorbs. Since hydrogen has only 1 shell, it forms the shortest covalent bond. That means a bond to hydrogen absorbs IR radiation with the highest wavenumber (2500–3500 cm^{-1}). Triple bonds between carbon atoms or between nitrogen and carbon atoms are the next shortest (2000–2500 cm^{-1}), followed by double bonds between carbon, nitrogen, or oxygen (1500–2000 cm^{-1}). Bonds that are part of a resonance system and therefore are in-between double and single bonds absorb radiation at frequency between that of double bonds and single bonds (1400–1500 cm^{-1}). Since oxygen and nitrogen are slightly smaller than carbon, a bond between carbon and oxygen or carbon and nitrogen has a slightly higher vibrational frequency than a comparable bond between carbon atoms.

To simplify the analysis of IR data, focus on bands above 1700 cm^{-1}. Use bands below 1600 cm^{-1} as needed to confirm known data, or to narrow down possibilities.

Electronegativity differences pull bond electrons toward the more electronegative atom, intensifying bond vibration. That is why the greater the electronegativity difference, the more pronounced the vibration, and therefore the more radiation that is absorbed.

Polarimetry:

Chiral compounds rotate plane-polarized light. Since each molecule of an enantiomer pair rotates light the same number of degrees, but in opposite directions, polarimetry can be used to determine the purity of a chiral compound.

UV-Visible Spectrophotometry:

The greater the amount of resonance, the lower the energy (the higher the wavelength) of UV-visible light a molecule absorbs. Since absorbance and concentration are proportional, the concentration of a sample can be calculated based on the amount of UV-Visible light it absorbs.

What You Need to Learn, Understand, and Apply

1. The types of information infrared spectroscopy provides. (page 432)
2. The general theory of infrared spectroscopy. (pages 433–434)
3. The ability to identify characteristic IR absorption bands for alcohols, amines, alkynes, alkenes, carboxylic acids, aldehydes, nitriles, carbonyl compounds, benzenes, esters, and ethers. (pages 434–442)
4. The ability to confirm IR absorption bands. (pages 442–443)
5. The ability to determine whether a molecule rotates plane-polarized light. (pages 443–445)
6. The ability to do calculations related to polarimetry data. (pages 445–446)
7. Working knowledge of the general theory and use of UV-Vis spectroscopy. (page 447)
8. The skills needed to apply the material and avoid common errors. (pages 448–449)

Chapter Preview

The general purpose of this chapter:

_____

_____

_____

_____

_____

(Keep this purpose in mind as you read the chapter to help you tie all the concepts together into one complete picture.)

Questions to Assess Understanding	Lecture/Reading After filling in your lecture note outline, fold the paper over so that only the assessment question is visible. Once you can consistently answer a given question correctly on your own, place an X by that question.
_____ What can be learned about a molecule from its IR data?	**Learning Objective 1: Know what types of information infrared spectrometry provides. (page 432)** Infrared (IR) spectrometry is used to determine: 1. _____ 2. _____
_____ What is the theory behind how IR spectrometry works?	**Learning Objective 2: Know the general theory of infrared spectrometry. (pages 433–434)** Explain the theory of how IR spectrometry works: _____ _____ _____ _____
_____ How does bond length correlate with vibrational frequency?	What is the relationship between the length of a bond and its vibrational frequency: _____
_____ What is one of the principle factors for determining the location of an IR band?	Because of that the location of an absorbance band is principally determined by: _____ _____ _____

_____ What is wavenumber and how does it correlate with frequency?

What wavenumber is and how it correlates with frequency: _____

_____

_____

_____ Where is the fingerprint region on an IR chart?

An IR chart is broken into sections known as the functional group region and the fingerprint region:

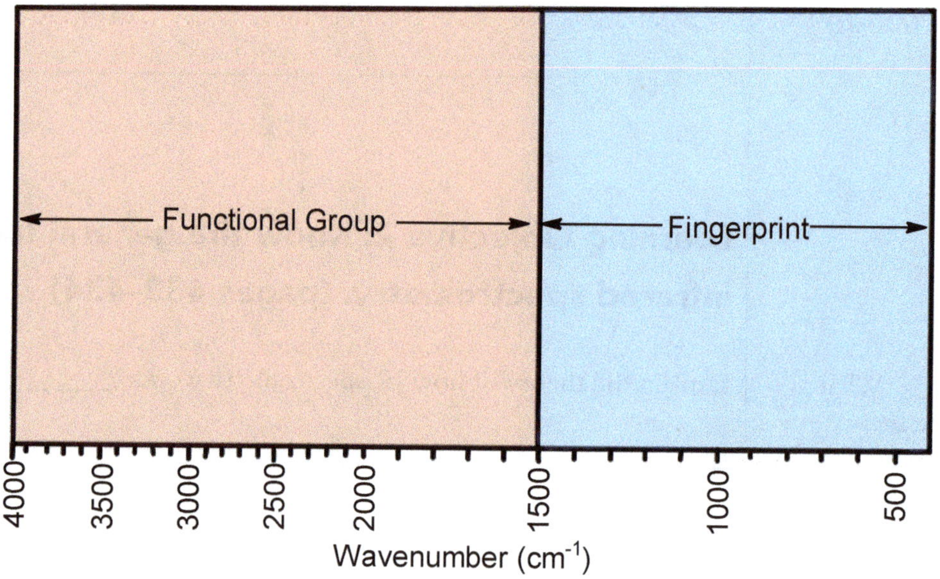

_____ What section of an IR chart provides the data you can interpret with the highest confidence?

Which region to prioritize and why: _____

_____

Learning Objective 3: Be able to identify unique IR absorption bands for alcohols, amines, alkynes, alkenes, carboxylic acids, aldehydes, nitriles, carbonyl compounds, benzene rings, esters, and ethers. (pages 434–442)

| ____ What types of bonds are found in the 2500 + cm^{-1} region of an IR?

____ What types of bonds are found in the 2000–2500 cm^{-1} region of an IR?

____ What types of bonds are found in the 1500–2000 cm^{-1} region of an IR?

____ What types of bonds are found in the 1400–1500 cm^{-1} region of an IR?

____ What specific type of bond is located in the general region of each of the following wavenumbers 3500, 3300, 3100, 2900, 2700, 2250, 2150, 1750, 1650, 1450, 1250, 1050?

____ What is an approximate wavenumber for the O–H bond of a carboxylic acid? | **General Type of Bond Corresponding to Each Color-Coded Region, and Specific Type of Bond Within Each Region:**

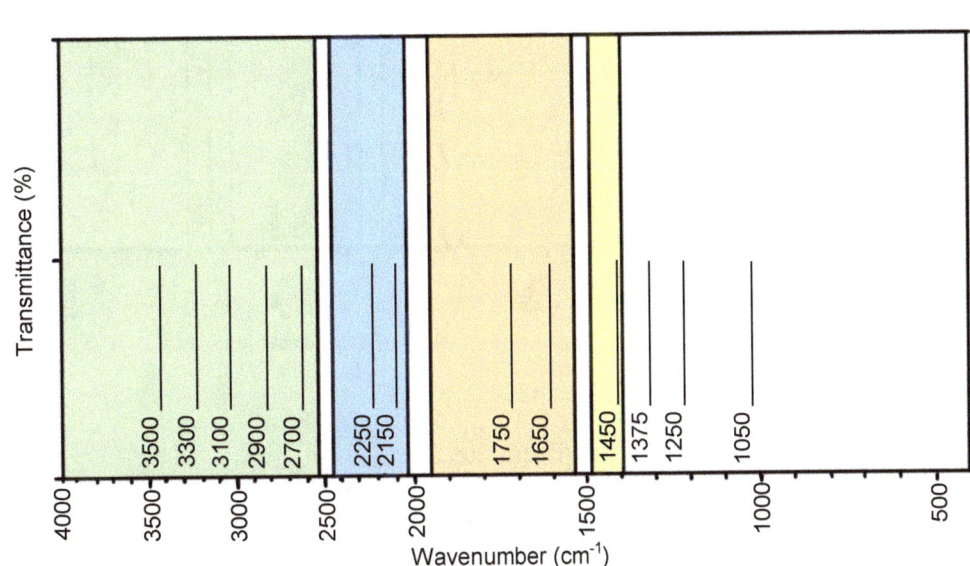

(The actual band will probably vary slightly from the wavenumbers shown above, so be flexible. The numbers shown above are meant to help you memorize the information. Notice that the numbers shown vary by 200 for the far left of the chart, and generally vary by 100 for the midrange values when breaks at 2500, 1950, and 1550 are factored in.)

Finer points related to interpreting IR bands:

1. **Hydrogen bonding weakens bonds and lowers vibrational frequency.**

     As a result, the absorbance band for an O–H bond of a carboxylic acid is shifted from approximately

     _____ for an alcohol OH to _____ |

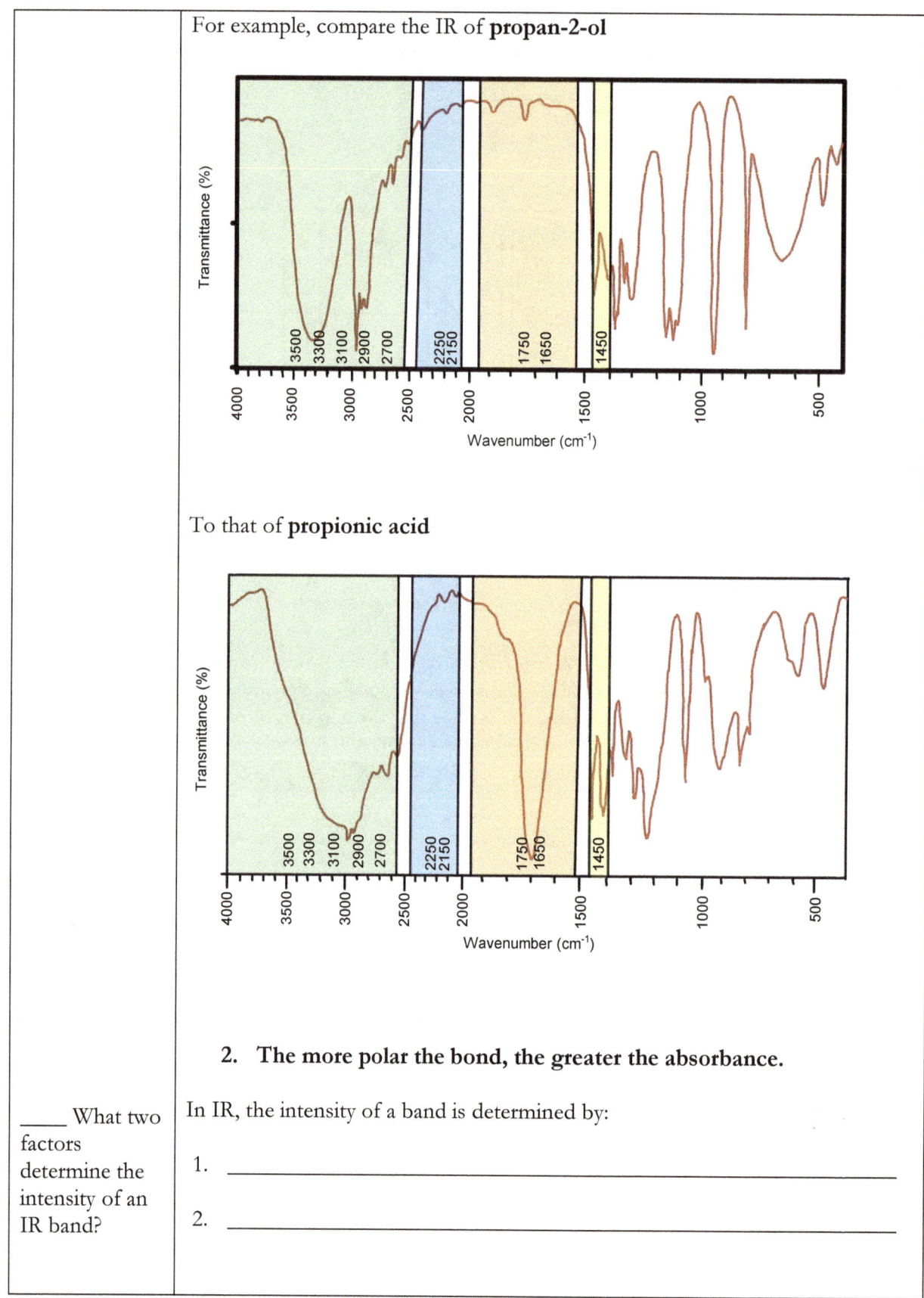

For example, compare the IR of **propan-2-ol**

To that of **propionic acid**

2. The more polar the bond, the greater the absorbance.

_____ What two factors determine the intensity of an IR band?

In IR, the intensity of a band is determined by:

1. _____

2. _____

This is particularly important to know when trying to distinguish between a bond to oxygen, and a comparable bond to nitrogen or carbon.

Example:

IR of aniline ($C_6H_5NH_2$)

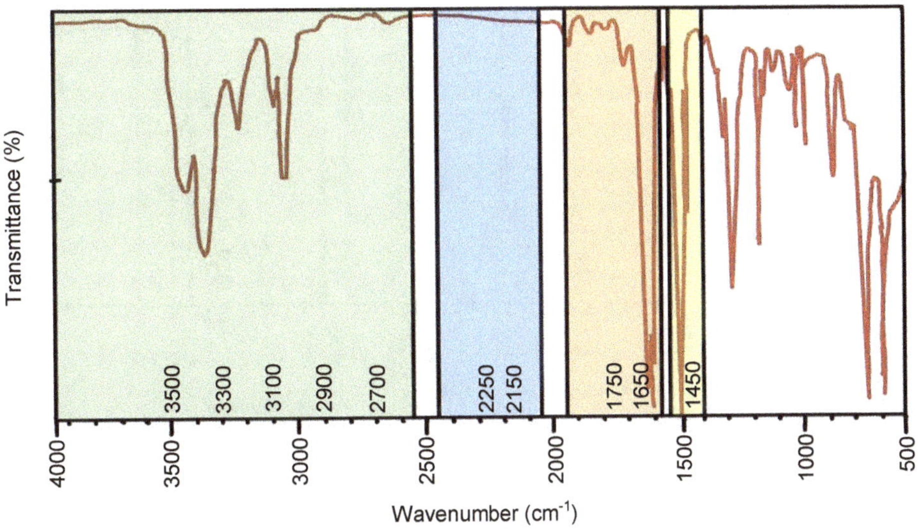

IR of propan-2-ol

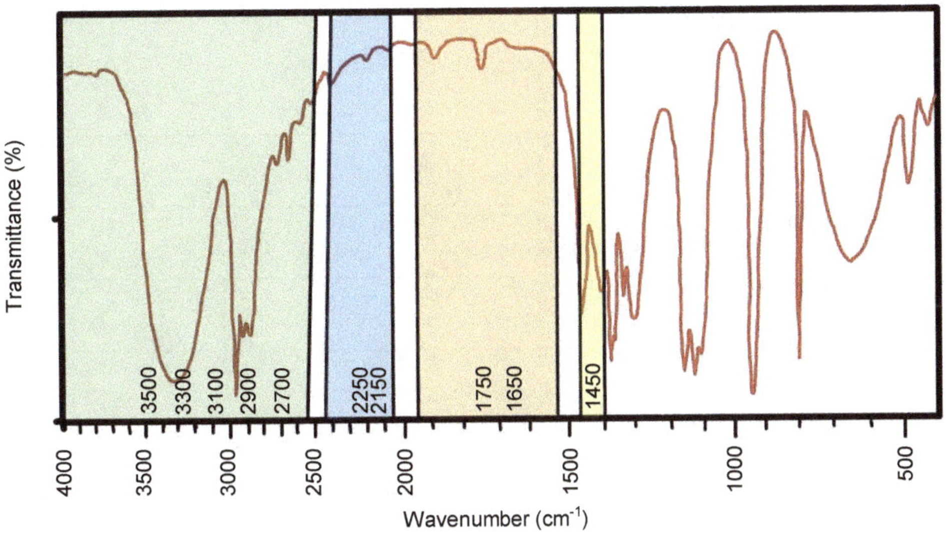

_____ What happens to an absorption band when a bond isn't polar?

What happens to an absorption band when a bond isn't polar _____

_____

Example:

IR of 2,3-dimethylbut-2-ene

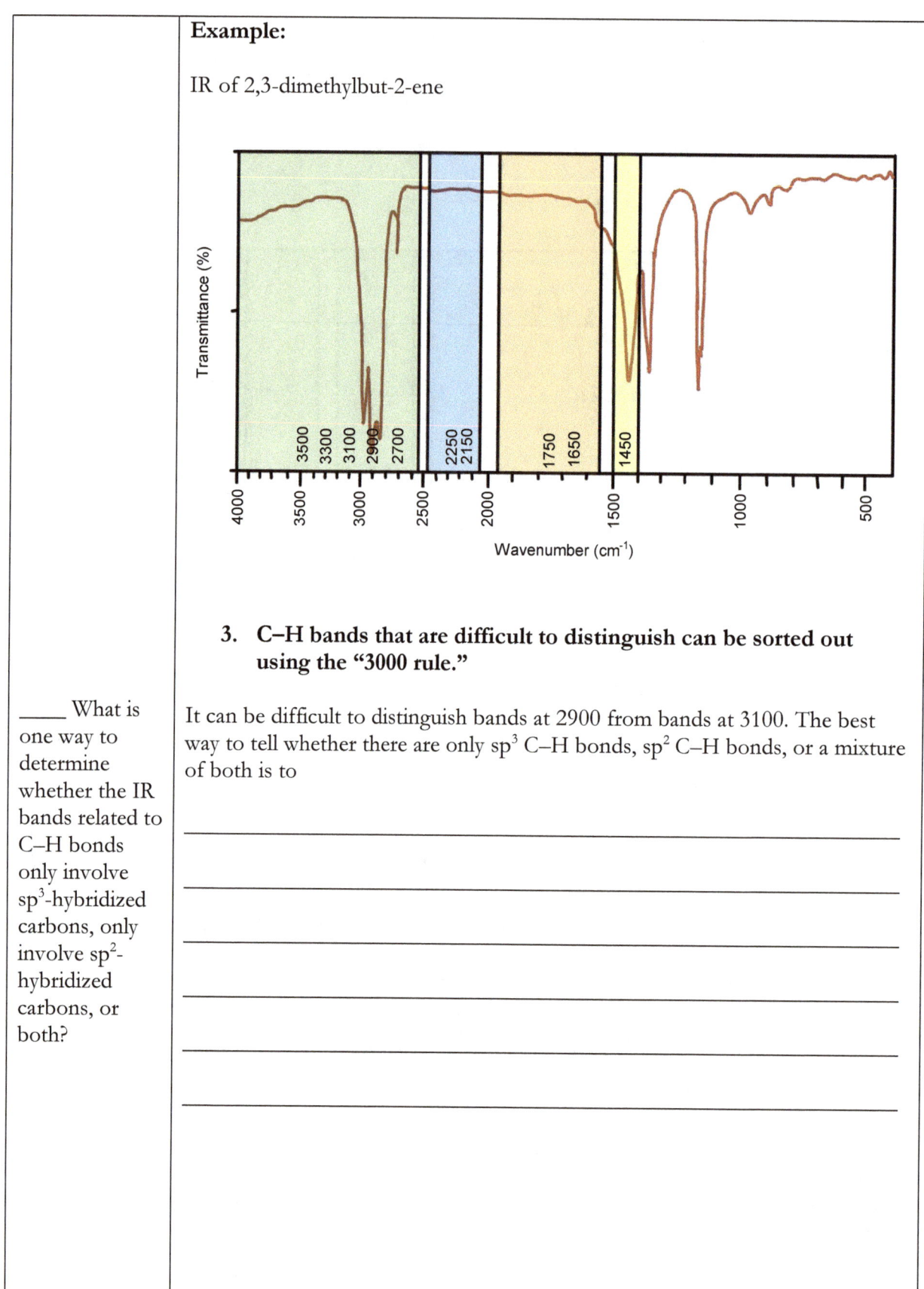

3. **C–H bands that are difficult to distinguish can be sorted out using the "3000 rule."**

It can be difficult to distinguish bands at 2900 from bands at 3100. The best way to tell whether there are only sp^3 C–H bonds, sp^2 C–H bonds, or a mixture of both is to

_____

_____

_____

_____

_____

_____

_____ What is one way to determine whether the IR bands related to C–H bonds only involve sp^3-hybridized carbons, only involve sp^2-hybridized carbons, or both?

Example:

1-phenylbutane

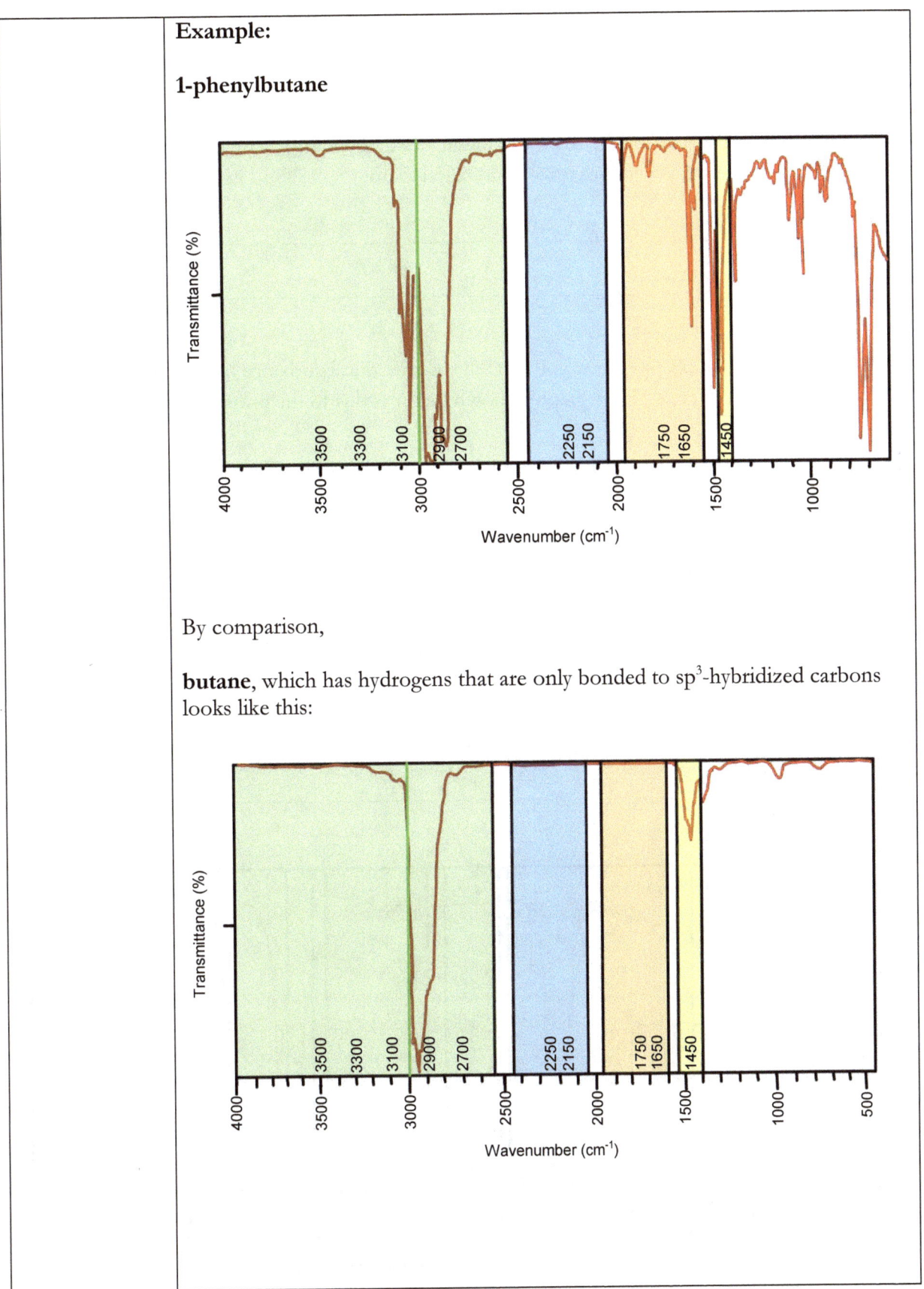

By comparison,

butane, which has hydrogens that are only bonded to sp³-hybridized carbons looks like this:

Learning Objective 4: Be able to confirm IR absorption bands. (pages 442–443)

_____ What is the best way to determine whether an ambiguous band is significant to your analysis?

The way to determine whether a weaker band or a band in the fingerprint region is significant to your analysis is to:

_____

_____

For example, how you can tell whether the absorbance band at approximately 1450 cm^{-1} in the graphic shown below is due to the presence of a benzene ring:

_____

_____

_____

How you can tell whether the absorbance band at approximately 2700 cm^{-1} in the graphic shown below is due to the presence of an aldehyde group:

_____

_____

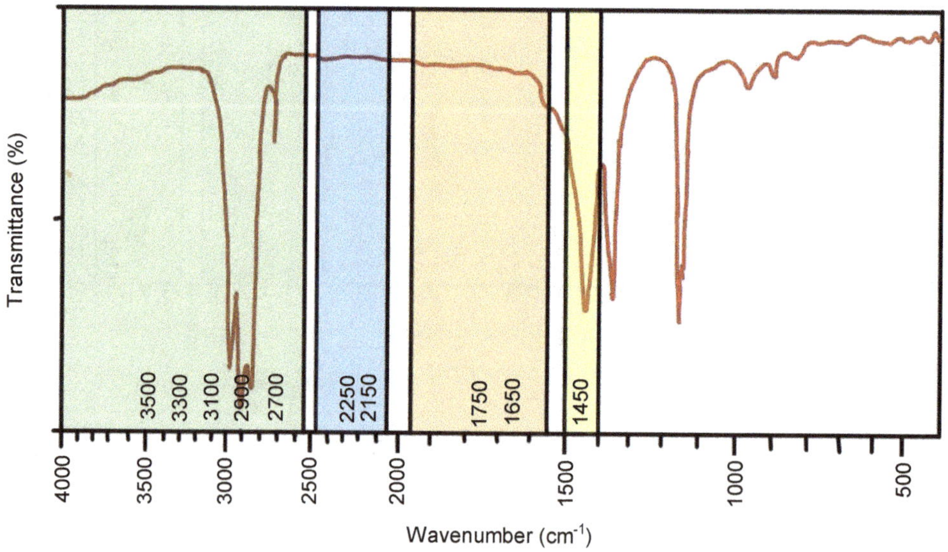

Learning Objective 5: Be able to determine whether any given organic compound can rotate plane-polarized light. (pages 443–445)

_____ How does a polarimeter work?

How a polarimeter works:

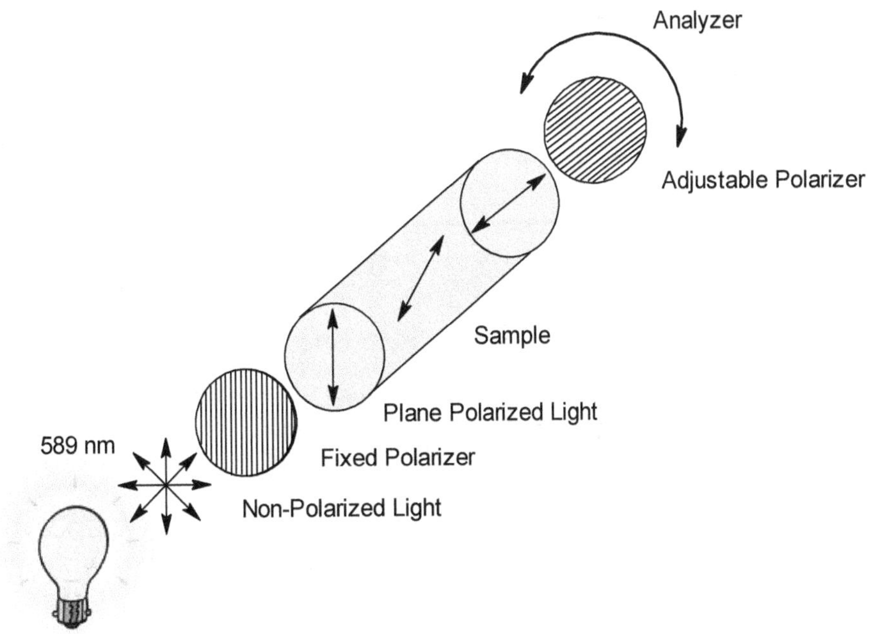

_____

_____

_____

_____

_____

Analyzer

Adjustable Polarizer

Sample

Plane Polarized Light

589 nm

Fixed Polarizer

Non-Polarized Light

_____ How is light rotated by a molecule compared to its enantiomer?

The way light is rotated by one molecule compared to its enantiomer.

_____

_____

___ What do the designations + and – mean when included in the name of a molecule?	What the designations + and – mean when included in the name of a molecule: + _____ – _____
___ What is the relationship between +/– and R/S?	The relationship between +/– and R/S _____ _____ _____
___ What types of molecules are optically active?	What types of molecules are optically active: _____ _____
___ What are 3 potential uses for polarimetry?	**Uses of Polarimetry** Polarimetry can be used to determine 1. _____ 2. _____ 3. _____ _____

Learning Objective 6: Know how to do calculations related to polarimetry. (pages 445–446)

____ What equation is used to calculate the specific optical rotation of a chiral substance (What is Biot's law)?

The equation used to calculate the specific optical rotation of a chiral substance (Biot's law):

What each of the symbols in the equation mean:

_____

_____

_____

_____

Calculating Optical Purity

____ What is the process for calculating the optical purity of a sample?

How to calculate the optical purity of a sample:

1. _____

_____

2. _____

_____

3. _____

_____

Use this information to calculate the percent of R and S stereoisomers in a solution if an unknown mixture has a specific rotation of -51.7°, and the specific rotation of the R form is -87.0 ° (and therefore the specific rotation of the S form is +87.0°).

Learning Objective 7: Know the general theory and use of UV-Vis spectroscopy. (page 447)

___ What is a characteristic of organic molecules that cause them to absorb light in the UV/Vis range, and, in general moves λ_{max} to higher wavelengths?

One characteristic of an organic molecule that causes it to absorb light in the UV/Vis range, and, in general, moves λ_{max} to higher wavelengths.

_____

_____

_____

___ What is the equation that is used to calculate the concentration of a substance from its absorbance? (What is the Beer-Lambert law?)

The equation that is used to calculate the concentration of a substance from its absorbance (Beer-Lambert law):

What each symbol in the equation means:

_____

_____

_____

_____

Learning Objective 8: Gain the skills needed to apply the material and avoid common errors. (pages 448–449)

____ What are a few items to keep in mind when ruling possibilities in/out based on an IR spectrum?

A few items to keep in mind when ruling possibilities in/out based on an IR spectrum:

_____

_____

_____

_____

_____

_____

____ What is the 5-step process to use when determining the identity of a molecule based on a combination of MS and IR data?

The process to use when determining the identity of a molecule based on a combination of MS and IR data:

1. _____

2. _____

3. _____

4. _____

_____

_____

5. _____

_____

_____

Learn to Analyze and Apply

Learn the Process of Determining the Identity of a Molecule Based on a Combination of MS and IR Data

Determine the structure of the molecule that produced the following MS and IR data:

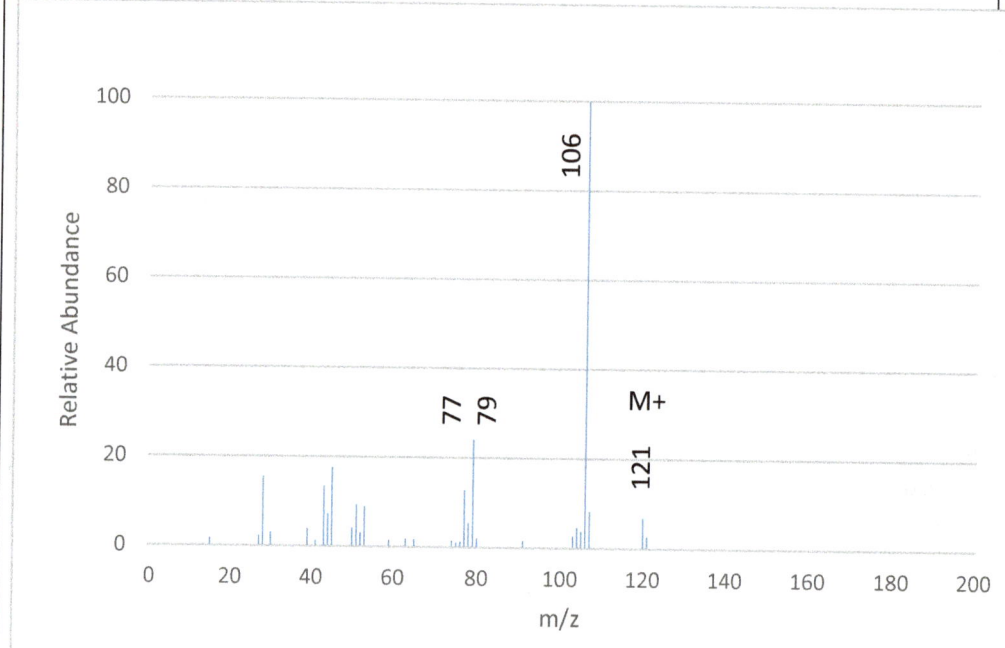

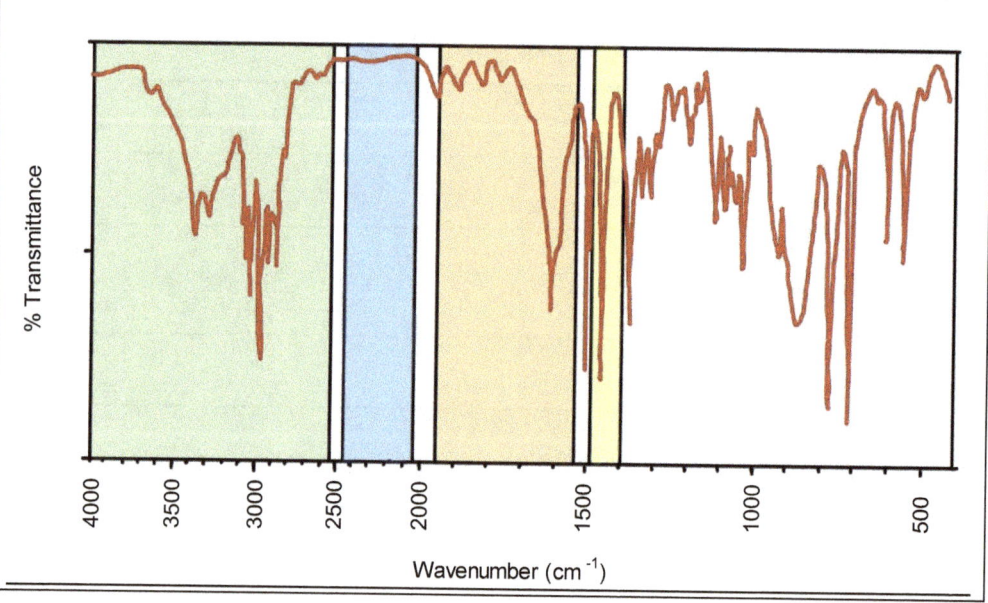

STEP 1:
Interpret the <mark>clearest MS data</mark>
(Look for evidence of Cl, Br, or N based on the M+/M+2 data.)

STEP 2:
Interpret the <mark>clearest IR data</mark>
Look for easily identifiable bands in the 1700–3500 cm⁻¹ range. When possible, verify your interpretation by looking for supporting bands.

STEP 3:
<mark>Verify agreement</mark> between MS and IR data

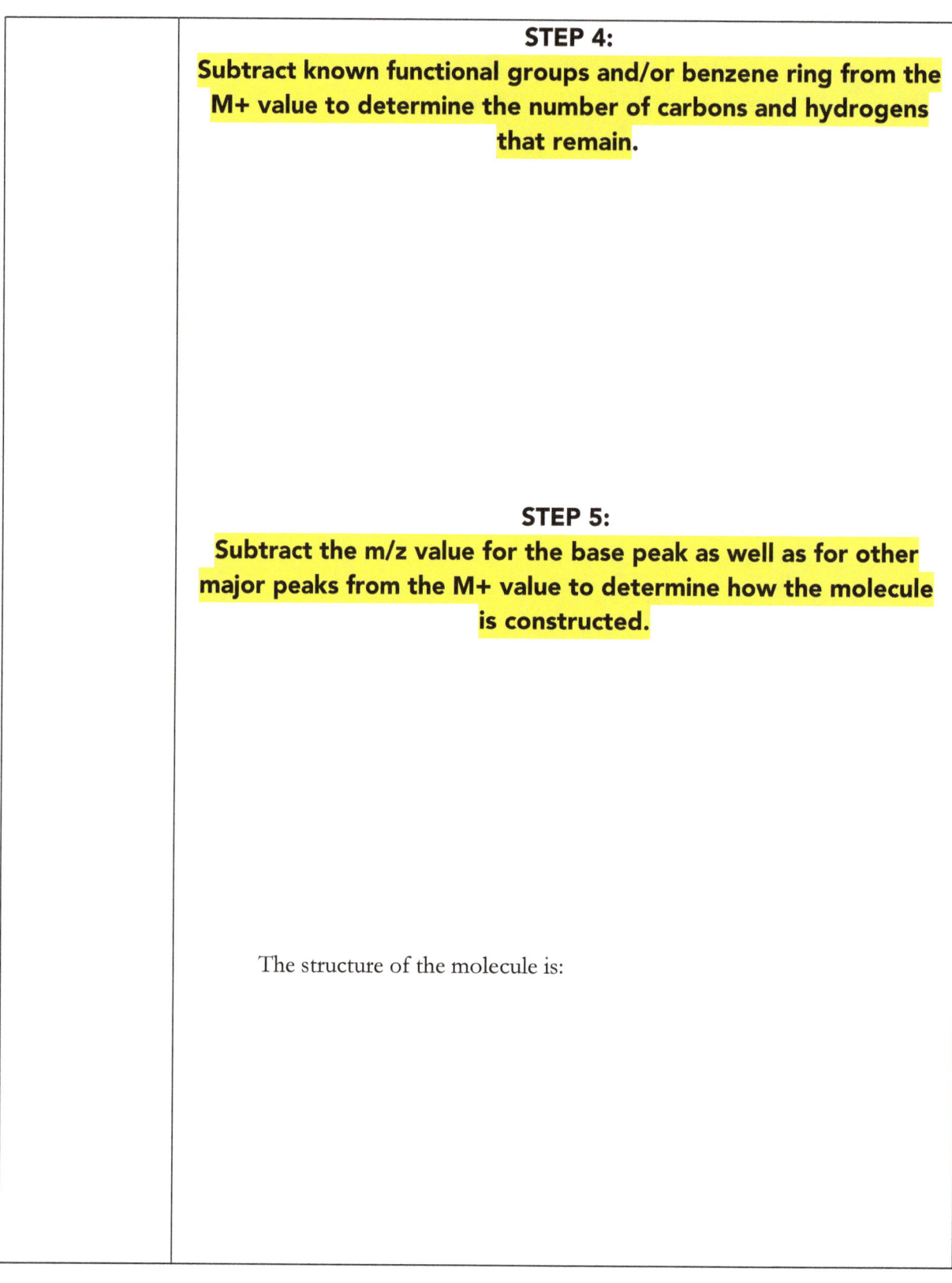

STEP 4:

==Subtract known functional groups and/or benzene ring from the M+ value to determine the number of carbons and hydrogens that remain.==

STEP 5:

==Subtract the m/z value for the base peak as well as for other major peaks from the M+ value to determine how the molecule is constructed.==

The structure of the molecule is:

Integrate Skills

Distinguish

1. Write the approximate wavenumbers you would expect to find in the IR spectrum for each of the following, and then list the wavenumber(s) that would help you distinguish between the molecules listed in each set.

    a. pent-1-yne vs. pent-2-yne

        pent-1-yne _____

        pent-2-yne _____

        How to distinguish _____

    b. butanoic acid vs. butan-1-ol

        butanoic acid _____

        butan-1-ol _____

        How to distinguish _____

    c. butanal vs. butanone

        butanal _____

        butanone _____

        How to distinguish _____

    d. benzoic acid vs. hexanoic acid

        benzoic acid _____

        hexanoic acid _____

        How to distinguish _____

e. ethanenitrile vs. ethyne

ethanenitrile _____

ethyne _____

How to distinguish _____

f. hex-2-ene vs. hex-2-yne

hex-2-ene _____

hex-2-yne _____

How to distinguish _____

g. butanone vs. methylpropanoate

butanone _____

methylpropanoate _____

How to distinguish _____

h. hexane vs. cyclohexane

hexane _____

cyclohexane _____

How to distinguish _____

i. butanal vs. butanoic acid

butanal _____

butanoic acid _____

How to distinguish _____

j. cyclohexene vs. benzene

cyclohexene _____

benzene _____

How to distinguish _____

Confirm

2. If you saw a band at approximately the indicated wavenumber, where else would you look to confirm the identity of the band?

a. 3300 _____

b. 3100 _____

c. 2700 _____

d. 3000 (broad, deep) _____

e. 1250 _____

Match to the Spectrum

3. The following IR spectra belong to the molecules acetonitrile, butanone, but-1-ene, but-1-yne, ethanol, 2-methylbutanal, and 1-phenylhexane. Write the structure and all characteristic bands you expect to see on an IR chart for each molecule. Then match each spectrum on the following pages to one of the molecules shown above.

Structure   Characteristic Bands

Acetonitrile _____   _____

Butanone _____   _____

But-1-ene _____   _____

But-1-yne _____   _____

Ethanol _____   _____

2-Methylbutanal _____   _____

1-Phenylhexane _____   _____

a. _____

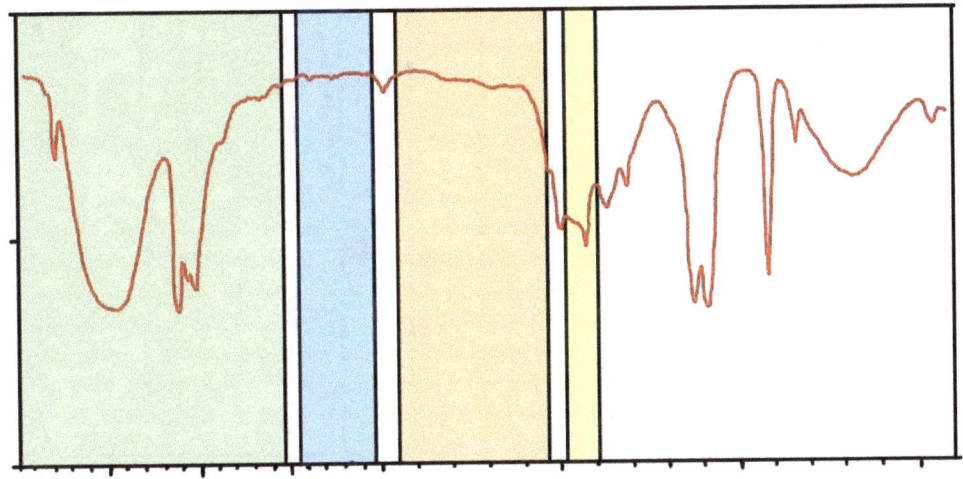

b. _____

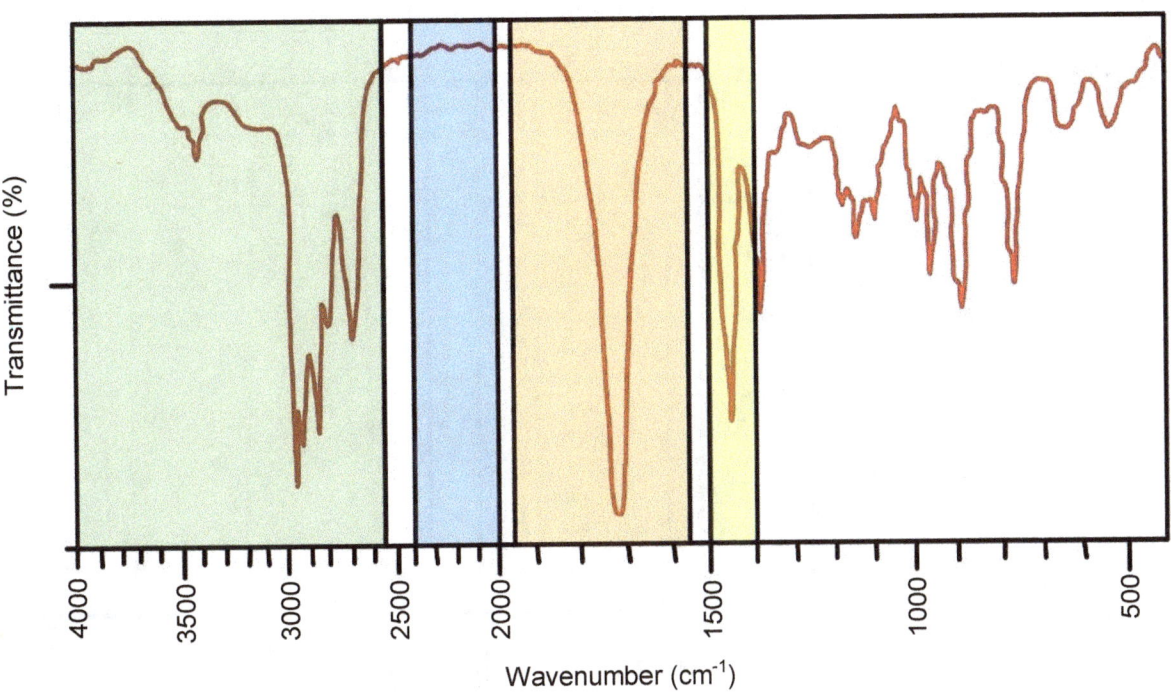

c.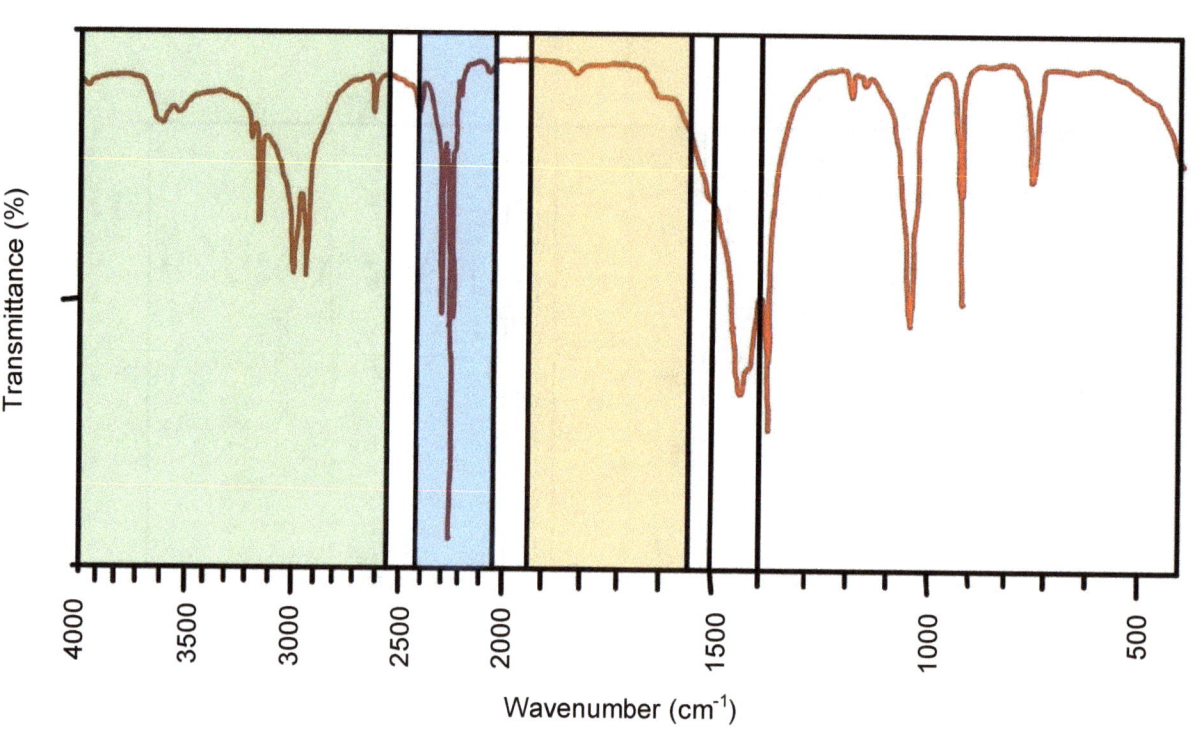

d.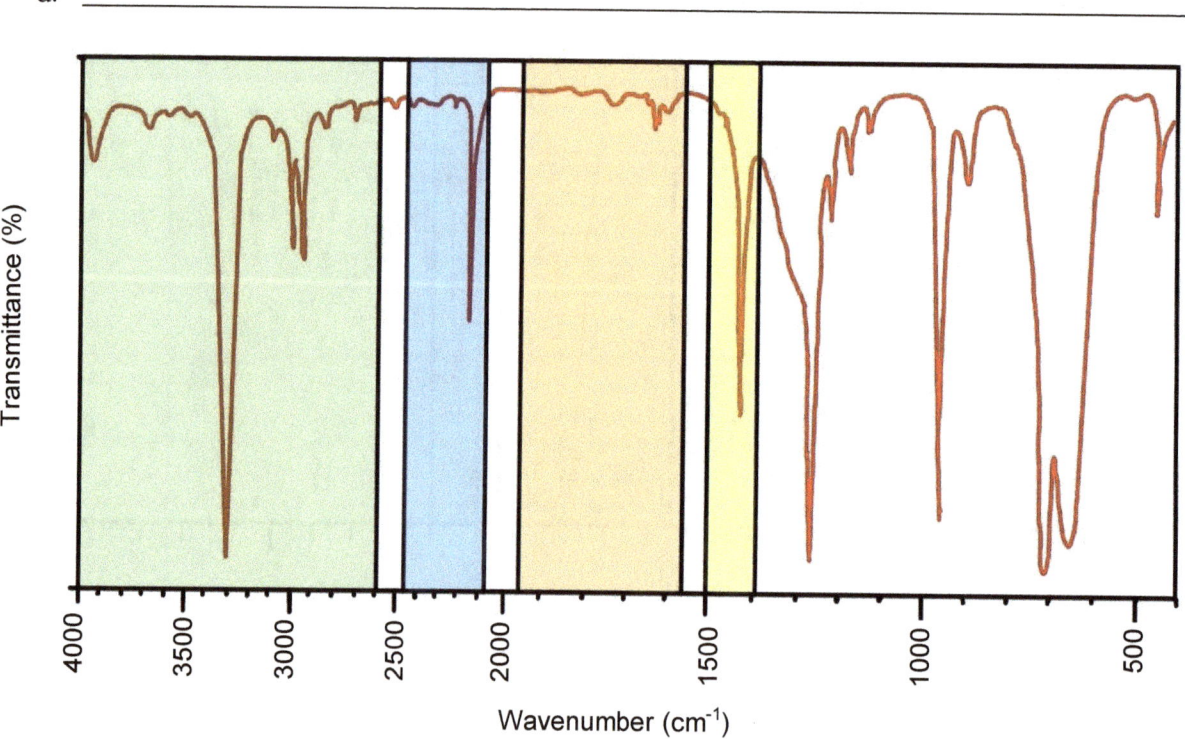

e. _____

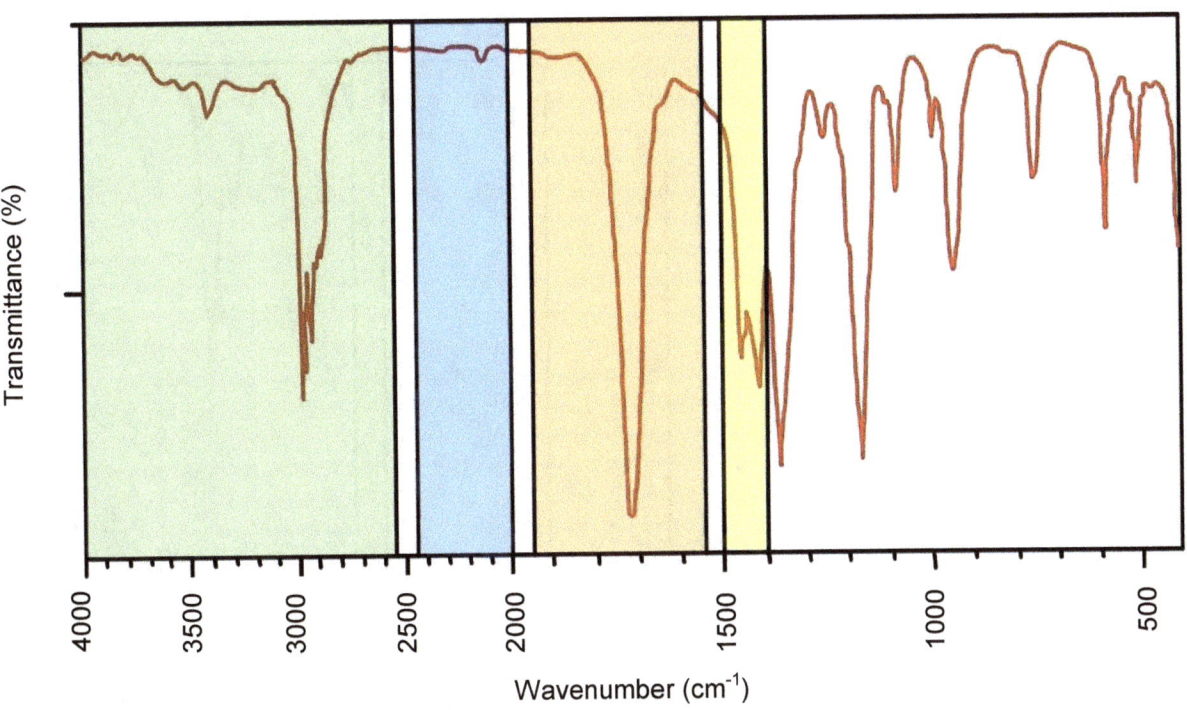

f. _____

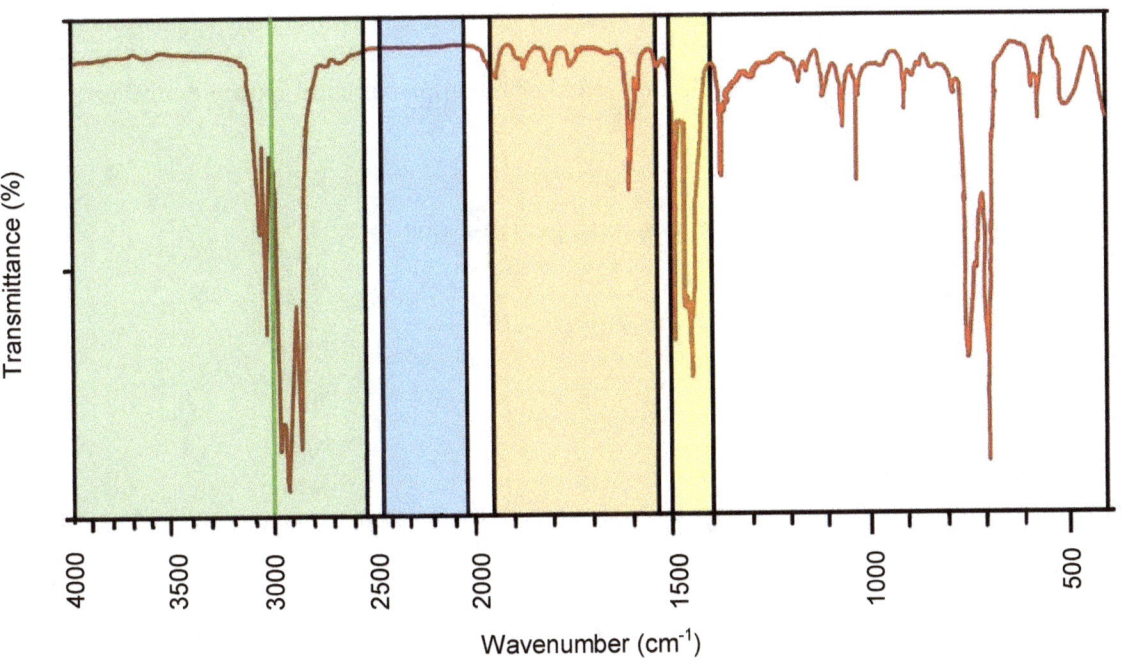

g. _____

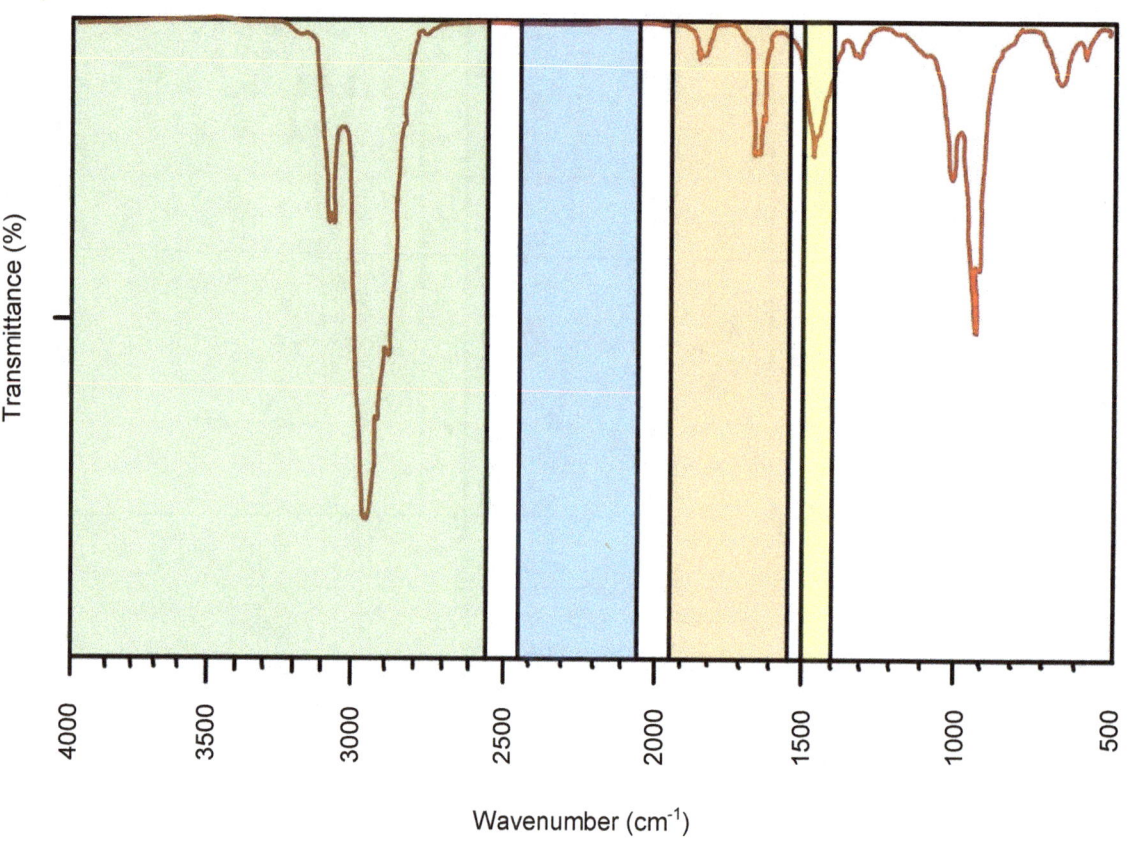

Match with Written Description

4. Within each set, select the molecule that matches the approximated IR wavenumbers:

a. 2950, 2750, 1740

   pentanoic acid, methyl pentanoate, pentanal, pentan-2-one

b. Multiple bands 3200–3000

   pent-1-ene, ethene, pent-1-yne, pent-2-yne, pentan-1-ol

c.   3350 (broad, deep), multiple bands 2850–2950

ethyl dimethyl amine, butan-1-amine, butan-1-ol, butanoic acid, but-1-yne

d.   Multiple bands 2850–2950, 1450

cyclohexane, benzene, toluene, benzaldehyde, cyclohexene, methyl cyclohexane

e.   3310 (narrow), multiple bands 2980–2880, 2120

ethyl dimethyl amine, butan-1-amine, butan-1-ol, butanoic acid, but-1-yne

f.   multiple bands 2949–2854, 1456, 1376

cyclohexane, benzene, toluene, benzaldehyde, cyclohexene, methyl cyclohexane

g. 3350 (narrow, shallow), multiple bands 3000–2850, 2120

   pent-1-ene, ethene, pent-1-yne, pent-2-yne, pentan-1-ol

h. 2876–2960, 1745, 1467, 1260, 1017

   pentanoic acid, methyl pentanoate, pentanal, pentan-2-one

i. Large bowl 2980 with overlay of peaks 2900, 1720 (broad, deep)

   ethyl dimethyl amine, butan-1-amine, butan-1-ol, butanoic acid, but-1-yne

j. 3100–2925, 1614, 1506, 1465, 1375

   cyclohexane, benzene, toluene, benzaldehyde, cyclohexene, methyl cyclohexane

k. Large bowl 2980 with overlay of peaks 2900, 1720 (broad, deep)

   pentanoic acid, methyl pentanoate, pentanal, pentan-2-one

l. Two narrow bands between 3300 and 3400, bands from 2850–2980

   ethyl dimethyl amine, butan-1-amine, butan-1-ol, butanoic acid, but-1-yne

m. Multiple bands 3073–2827, 2745. 1696 (broad, deep), 1456

   cyclohexane, benzene, toluene, benzaldehyde, cyclohexene, methyl cyclohexane

n. Multiple bands 2850–3000, 1717 (broad, deep)

   pentanoic acid, methyl pentanoate, pentanal, pentan-2-one

o. Multiple bands 3082–3009, 1630, 1496

ethylbenzene, styrene, cyclohexane, hex-2-ene

p. Multiple bands 3030–2820, 1655 (narrow, shallow), 1437

cyclohexane, benzene, toluene, benzaldehyde, cyclohexene, methyl cyclohexane

q. Multiple bands 3000–2800, 1450

ethyl dimethyl amine, butan-1-amine, butan-1-ol, butanoic acid, but-1-yne

Polarimetry Calculations

5. If a pure sample of the R-form of an enantiomer has a specific rotation of -34 degrees, what is the proportion of R and S if a mixture has a specific rotation of + 20 degrees?

Use of the Beer-Lambert Law

6. If a solution containing a protein with a molar absorptivity (extinction coefficient) value of 140,000 M^{-1} cm^{-1} at 280 nm has an absorbance of 1.27 with a 1 cm path length, what is the concentration of that protein in the solution?

Mash-up 1

1. (Chapters 1 and 2) Assign an IUPAC name to the following:

Stereochemistry

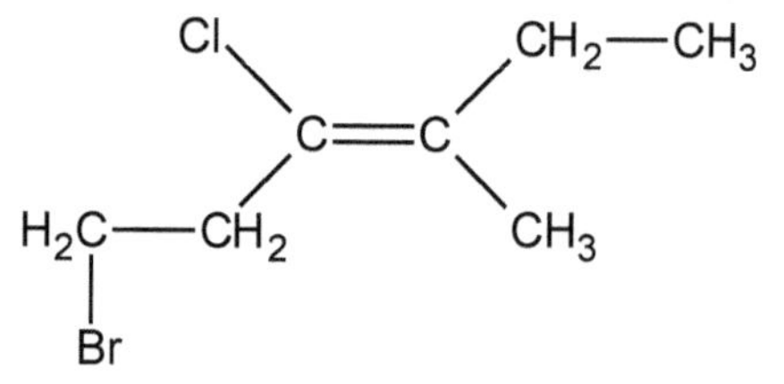

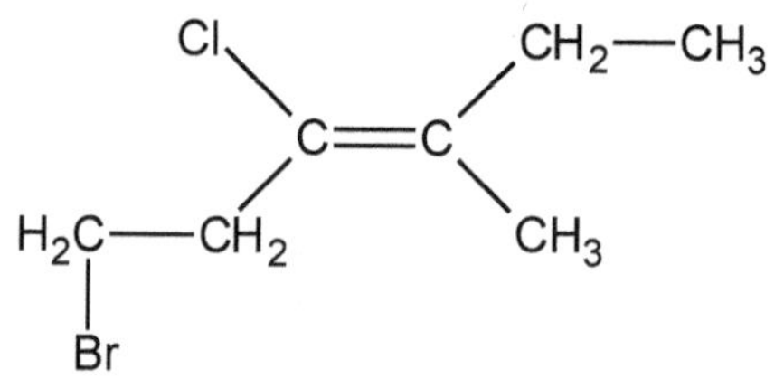

_____

2. (Chapters 1 and 3) Assign both an IUPAC name and a common name to the following:

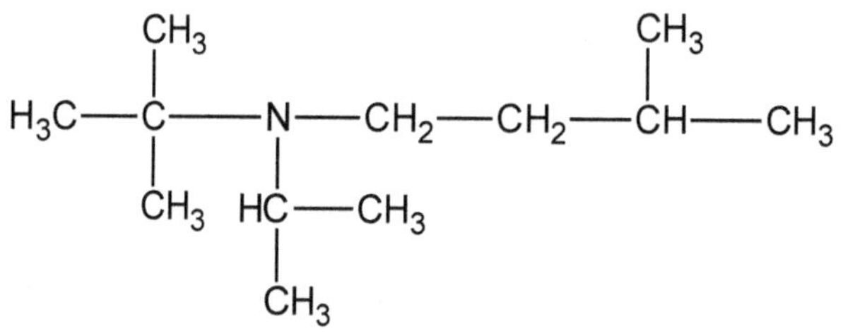

IUPAC Name: _____

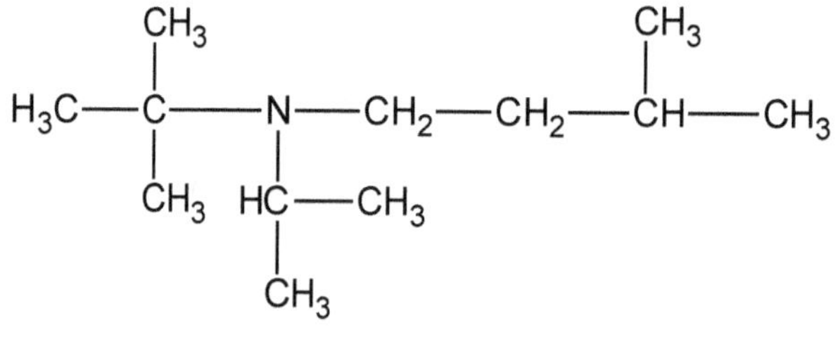

Common Name _____

3. (Chapters 1 and 3) Assign both an IUPAC name and a common name to the following:

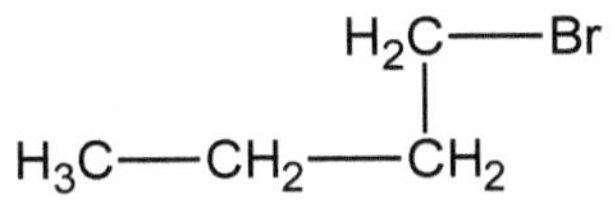

H₂C—Br
H₃C—CH₂—CH₂

(structure)

IUPAC Name _____

(structure)

H₂C—Br
H₃C—CH₂—CH₂

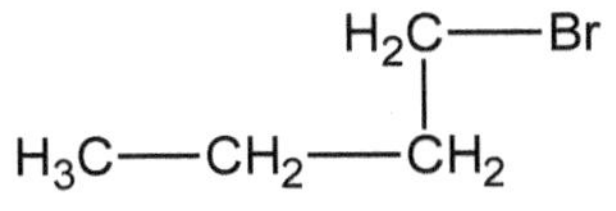

Common Name _____

4. (Chapter 10) Assign a common name to the following:

H₂C—CH₃

(benzene ring structure)

CH=CH₂

_____

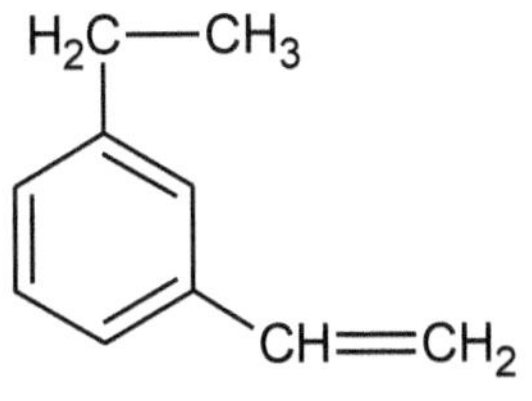

5. (Chapter 10) Assign a common name to the following:

H₃C—CH₂—CH₂—CH—CH₃
                |
               CH₂
                |
         (benzene ring)

_____

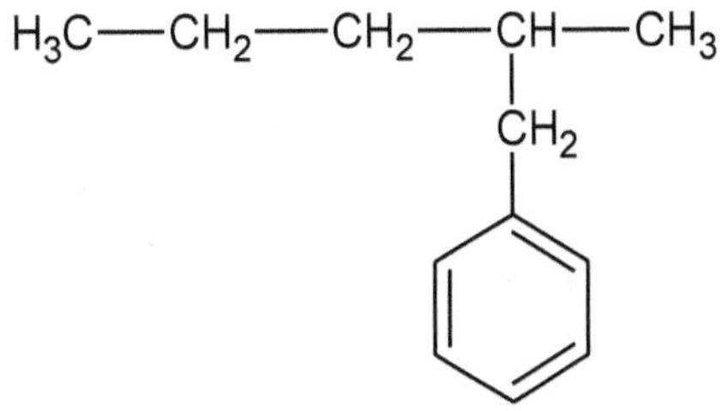

6. (Chapter 11) Write the correct IUPAC name and common name for the following:

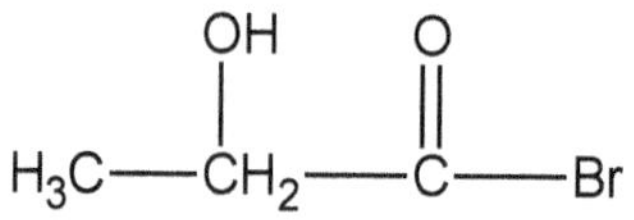

$$H_3C-CH_2-\overset{\overset{\displaystyle OH}{|}}{C}\overset{\overset{\displaystyle O}{\|}}{}{-}Br$$

IUPAC Name: _____

$$H_3C-CH_2-\overset{\overset{\displaystyle OH}{|}}{C}\overset{\overset{\displaystyle O}{\|}}{}{-}Br$$

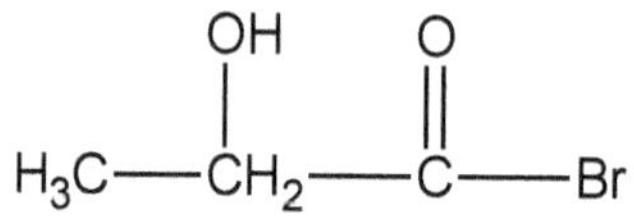

Common Name _____

7. (Chapter 11) Write the correct common name and derived name for the following:

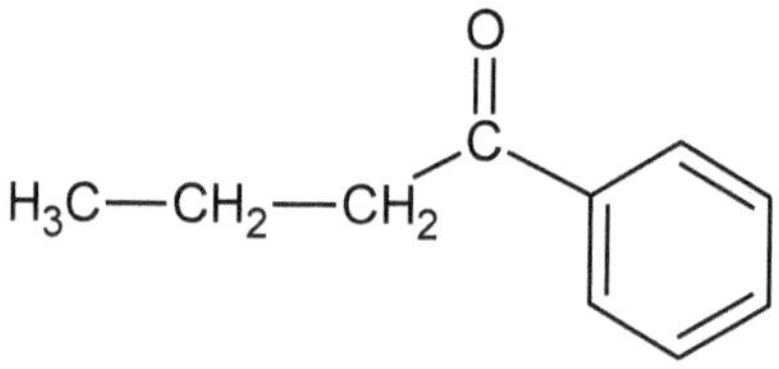

$$H_3C-CH_2-CH_2-\overset{\overset{\displaystyle O}{\|}}{C}-C_6H_5$$

Common Name _____

$$H_3C-CH_2-CH_2-\overset{\overset{\displaystyle O}{\|}}{C}-C_6H_5$$

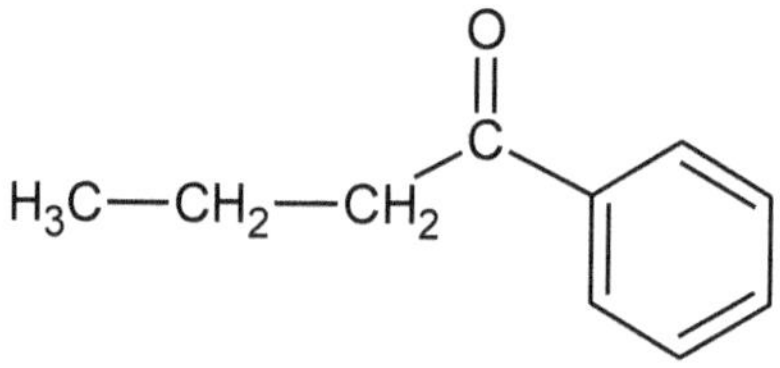

Derived Name _____

8. (Chapter 13) Write two correct IUPAC names for the following:

IUPAC Name 1: _____

IUPAC Name 2: _____

9. (Chapters 9 and 10) Which of the following is more stable and why?

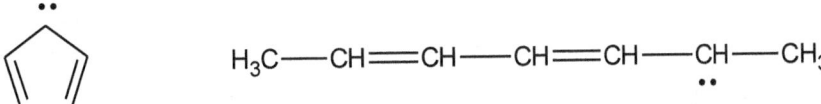

_____

_____

10. (Chapter 12) What type of instability will need to be addressed in the intermediate that is created when the carbon adjacent to a carbonyl carbon reacts with a nucleophile/base?

_____

11. (Chapter 7) What type of instability will need to be addressed in the intermediate that is created when a pi bond reacts with an electrophilic atom that has a non-bonded electron pair?

_____

12. (Chapter 13) What general resources could be used to bring electrons to a carbon that has a partial positive charge?

_____

_____

_____

13. (Chapters 6 and 8) What general methods could potentially be used to stabilize a carbon radical?

_____

_____

(Chapters 5–13) For each of the following,

14. a. Label the reactive features, highlight the most reactive feature, and then highlight what it needs. Also, state if a carbocation, carbon radical, or carbanion will start to develop, and/or if aromatic character will be lost because of a reaction between these molecules. If a carbocation, carbon radical or carbanion starts to develop, label where that will occur.

methyl acetate + sodium methoxide in DMSO

b. Use mechanism arrows to illustrate the reaction that occurs.

If applicable, use stabilization resources to deal with the carbocation, carbon radical, or carbanion that starts to develop during the reaction, and draw the structure of any resonance-stabilized intermediate.

Continue labelling and diagramming the reaction until you find the major stable product(s).

Finally, state the stereochemistry of the major product(s) and use either Fisher projection or perspective formula representations to illustrate that stereochemistry.

15. a. Label the reactive features, highlight the most reactive feature, and then highlight what it needs. Also, state if a carbocation, carbon radical, or carbanion will start to develop, and/or if aromatic character will be lost because of a reaction between these molecules. If a carbocation, carbon radical or carbanion starts to develop, label where that will occur.

2,3-dimethylpenta-1,3-diene with Cl_2

b. Use mechanism arrows to illustrate the reaction that occurs.

If applicable, use stabilization resources to deal with the carbocation, carbon radical, or carbanion that starts to develop during the reaction, and draw the structure of any resonance-stabilized intermediate.

Continue labelling and diagramming the reaction until you find the major stable product(s).

Finally, state the stereochemistry of the major product(s) and use either Fisher projection or perspective formula representations to illustrate that stereochemistry.

16. a. Label the reactive features, highlight the most reactive feature, and then highlight what it needs. Also, state if a carbocation, carbon radical, or carbanion will start to develop, and/or if aromatic character will be lost because of a reaction between these molecules. If a carbocation, carbon radical or carbanion starts to develop, label where that will occur.

+ Cl₂ and FeCl₃

b. Use mechanism arrows to illustrate the reaction that occurs.

If applicable, use stabilization resources to deal with the carbocation, carbon radical, or carbanion that starts to develop during the reaction, and draw the structure of any resonance-stabilized intermediate.

Continue labelling and diagramming the reaction until you find the major stable product(s).

Finally, state the stereochemistry of the major product(s) and use either Fisher projection or perspective formula representations to illustrate that stereochemistry.

17. a. Label the reactive features, highlight the most reactive feature, and then highlight what it needs. Also, state if a carbocation, carbon radical, or carbanion will start to develop, and/or if aromatic character will be lost because of a reaction between these molecules. If a carbocation, carbon radical or carbanion starts to develop, label where that will occur.

acetone with methylamine

b. Use mechanism arrows to illustrate the reaction that occurs.

If applicable, use stabilization resources to deal with the carbocation, carbon radical, or carbanion that starts to develop during the reaction, and draw the structure of any resonance-stabilized intermediate.

Continue labelling and diagramming the reaction until you find the major stable product(s).

Finally, state the stereochemistry of the major product(s) and use either Fisher projection or perspective formula representations to illustrate that stereochemistry.

18. a. Label the reactive features, highlight the most reactive feature, and then highlight what it needs. Also, state if a carbocation, carbon radical, or carbanion will start to develop, and/or if aromatic character will be lost because of a reaction between these molecules. If a carbocation, carbon radical or carbanion starts to develop, label where that will occur.

3,3-dimethylhept-1-ene + Cl_2 in CH_2Cl_2

b. Use mechanism arrows to illustrate the reaction that occurs.

If applicable, use stabilization resources to deal with the carbocation, carbon radical, or carbanion that starts to develop during the reaction, and draw the structure of any resonance-stabilized intermediate.

Continue labelling and diagramming the reaction until you find the major stable product(s).

Finally, state the stereochemistry of the major product(s) and use either Fisher projection or perspective formula representations to illustrate that stereochemistry.

19. a. Label the reactive features, highlight the most reactive feature, and then highlight what it needs. Also, state if a carbocation, carbon radical, or carbanion will start to develop, and/or if aromatic character will be lost because of a reaction between these molecules. If a carbocation, carbon radical or carbanion starts to develop, label where that will occur.

    (R)-butan-2-ol with $POCl_3$ in pyridine

    b. Use mechanism arrows to illustrate the reaction that occurs.

    If applicable, use stabilization resources to deal with the carbocation, carbon radical, or carbanion that starts to develop during the reaction, and draw the structure of any resonance-stabilized intermediate.

    Continue labelling and diagramming the reaction until you find the major stable product(s).

    Finally, state the stereochemistry of the major product(s) and use either Fisher projection or perspective formula representations to illustrate that stereochemistry.

20. a. Label the reactive features, highlight the most reactive feature, and then highlight what it needs. Also, state if a carbocation, carbon radical, or carbanion will start to develop, and/or if aromatic character will be lost because of a reaction between these molecules. If a carbocation, carbon radical or carbanion starts to develop, label where that will occur.

1-methylcyclopenta-1,3-diene with $CH_2=CHCH=O$

b. Use mechanism arrows to illustrate the reaction that occurs.

If applicable, use stabilization resources to deal with the carbocation, carbon radical, or carbanion that starts to develop during the reaction, and draw the structure of any resonance-stabilized intermediate.

Continue labelling and diagramming the reaction until you find the major stable product(s).

Finally, state the stereochemistry of the major product(s) and use either Fisher projection or perspective formula representations to illustrate that stereochemistry.

21. a. Label the reactive features, highlight the most reactive feature, and then highlight what it needs. Also, state if a carbocation, carbon radical, or carbanion will start to develop, and/or if aromatic character will be lost because of a reaction between these molecules. If a carbocation, carbon radical or carbanion starts to develop, label where that will occur.

+ HNO_3 with tr. H_2SO_4

b. Use mechanism arrows to illustrate the reaction that occurs.

If applicable, use stabilization resources to deal with the carbocation, carbon radical, or carbanion that starts to develop during the reaction, and draw the structure of any resonance-stabilized intermediate.

Continue labelling and diagramming the reaction until you find the major stable product(s).

Finally, state the stereochemistry of the major product(s) and use either Fisher projection or perspective formula representations to illustrate that stereochemistry.

22. a. Label the reactive features, highlight the most reactive feature, and then highlight what it needs. Also, state if a carbocation, carbon radical, or carbanion will start to develop, and/or if aromatic character will be lost because of a reaction between these molecules. If a carbocation, carbon radical or carbanion starts to develop, label where that will occur.

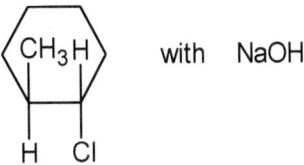   with   NaOH

b. Use mechanism arrows to illustrate the reaction that occurs.

If applicable, use stabilization resources to deal with the carbocation, carbon radical, or carbanion that starts to develop during the reaction, and draw the structure of any resonance-stabilized intermediate.

Continue labelling and diagramming the reaction until you find the major stable product(s).

Finally, state the stereochemistry of the major product(s) and use either Fisher projection or perspective formula representations to illustrate that stereochemistry.

23. a. Label the reactive features, highlight the most reactive feature, and then highlight what it needs. Also, state if a carbocation, carbon radical, or carbanion will start to develop, and/or if aromatic character will be lost because of a reaction between these molecules. If a carbocation, carbon radical or carbanion starts to develop, label where that will occur.

m-fluoropropylbenzene + HNO_3 with tr. H_2SO_4

b. Use mechanism arrows to illustrate the reaction that occurs.

If applicable, use stabilization resources to deal with the carbocation, carbon radical, or carbanion that starts to develop during the reaction, and draw the structure of any resonance-stabilized intermediate.

Continue labelling and diagramming the reaction until you find the major stable product(s).

Finally, state the stereochemistry of the major product(s) and use either Fisher projection or perspective formula representations to illustrate that stereochemistry.

24. A substance with the following MS and IR spectra is reacted with 2-methylbuta-1,3-diene. Draw the structure of the unknown compound and then draw the product of the reaction.

MS:

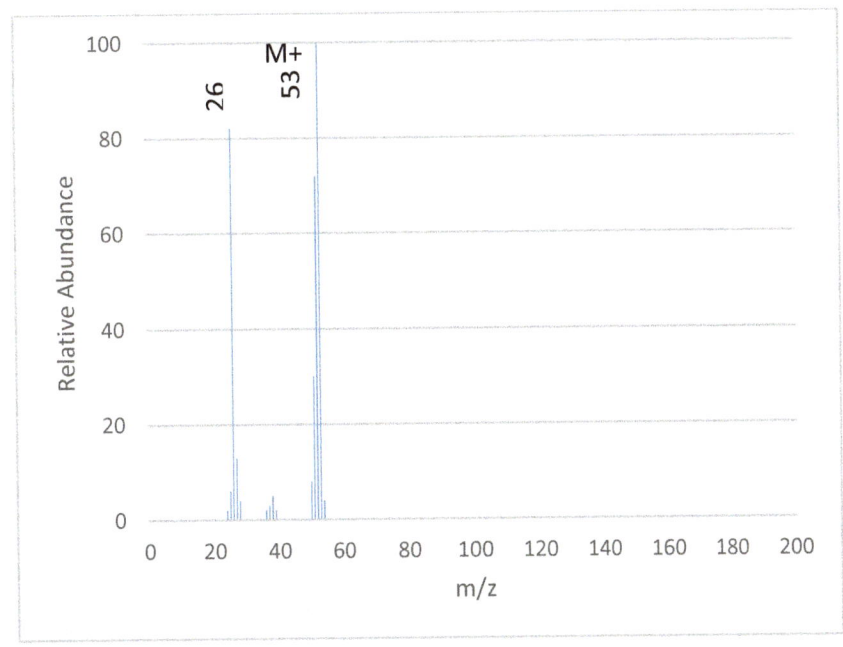

IR:

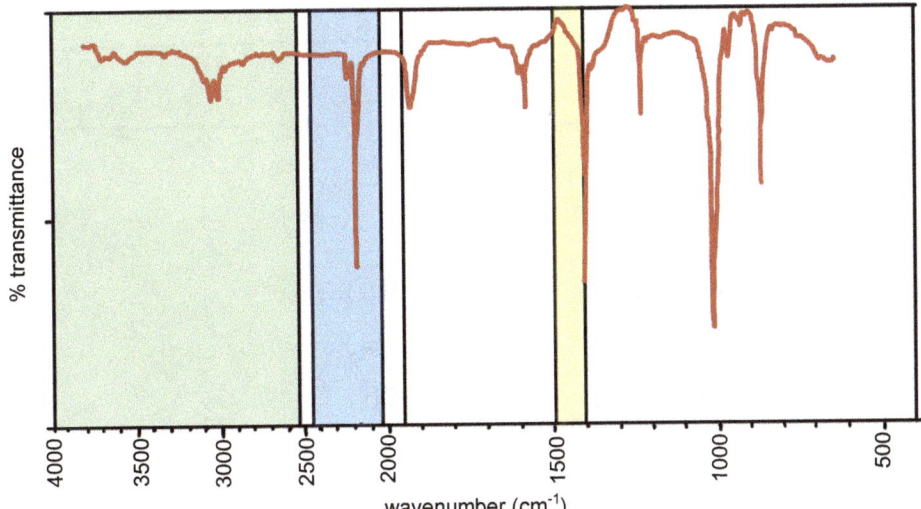

_____   _____

Mash-up 2

1. (Chapter 3) Assign a common name to the following:

$$H_3C-\underset{\underset{CH_3}{|}}{\overset{\overset{CH_3}{|}}{C}}-CH_2-O-\underset{\underset{CH_3}{|}}{\overset{\overset{CH_3}{|}}{C}}-CH_3$$

_____

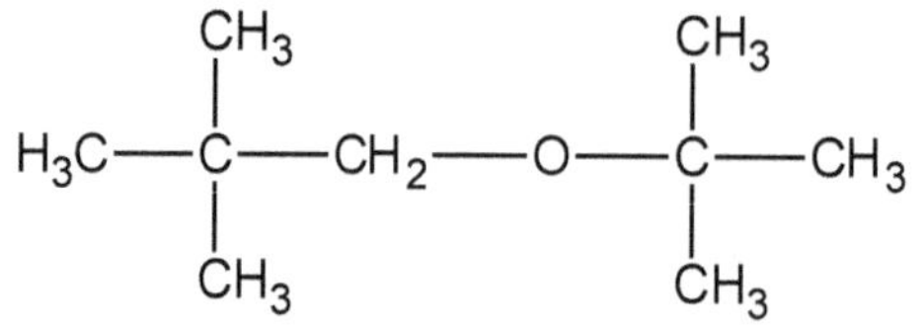

2. (Chapters 1 and 3) Assign both an IUPAC name and a common name to the following:

$$H_2C=CH-CH_2-Br$$

IUPAC Name: _____

$$H_2C=CH-CH_2-Br$$

Common Name: _____

3. (Chapter 4) Assign a mixed IUPAC name to the following:

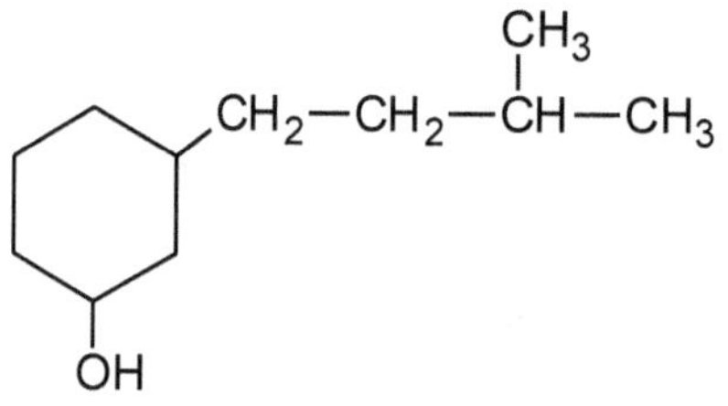

_____

4. (Chapter 10) Assign a common name to the following:

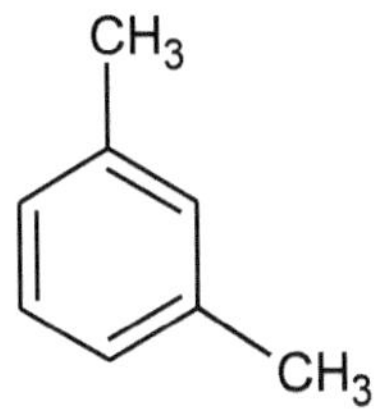

____

5. (Chapter 10) Assign a common name to the following:

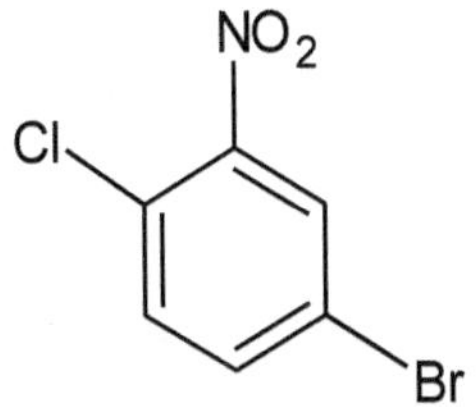

____

6. (Chapter 11) Write the correct IUPAC name and common name for the following:

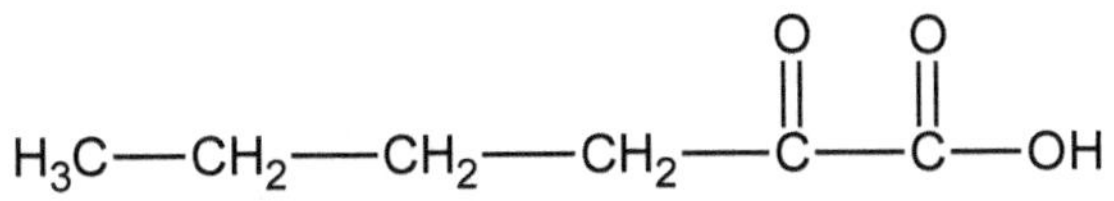

IUPAC Name: ______________________________

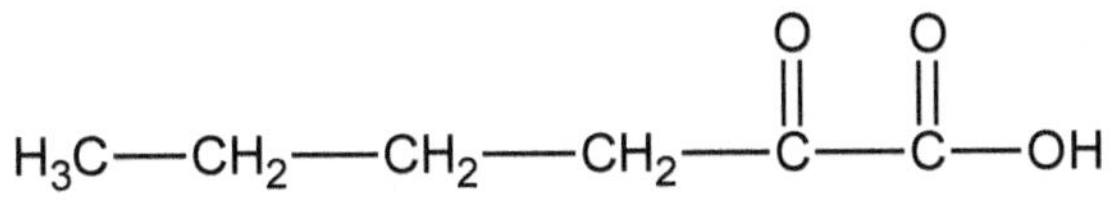

Common Name ______________________________

7. (Chapter 11) Write the correct IUPAC name and common name for the following:

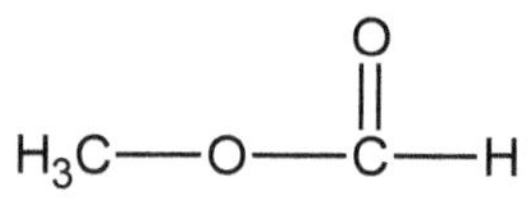

$$H_3C\!-\!O\!-\!\overset{\displaystyle O}{\overset{\|}{C}}\!-\!H$$

IUPAC Name: _____

$$H_3C\!-\!O\!-\!\overset{\displaystyle O}{\overset{\|}{C}}\!-\!H$$

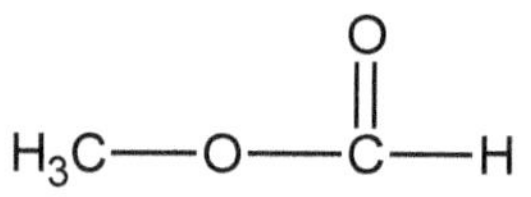

Common Name _____

8. (Chapter 13) Write two correct IUPAC names and the common name for the following:

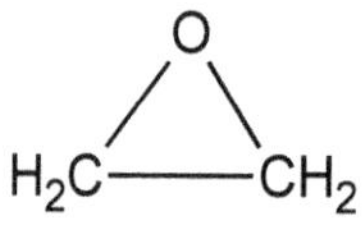

$$H_2C\!-\!\!-\!\!-\!CH_2 \quad (O)$$

IUPAC Name: _____

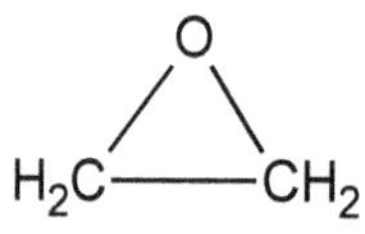

$$H_2C\!-\!\!-\!\!-\!CH_2 \quad (O)$$

IUPAC Name: _____

$$H_2C\!-\!\!-\!\!-\!CH_2 \quad (O)$$

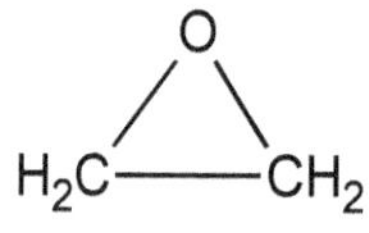

Common Name _____

9. (Chapter 12) What type of instability will need to be addressed in the intermediate that is created when a carbonyl carbon reacts with a nucleophile/base?

_____

10. (Chapter 9) What type of instability will need to be addressed in the intermediate that is created when an enol reacts under acidic conditions?

_____

11. (Chapter 10) What type of instability will need to be addressed in the intermediate when an aromatic compound reacts with a very reactive electrophile?

_____

12. (Chapter 9) What general resource could potentially be used to stabilize a carbocation when aromatic character needs to be restored?

_____

_____

_____

_____

_____

13. (Chapter 9) What general methods could potentially be used to stabilize a carbanion?

_____

_____

_____

(Chapters 5–13) For each of the following,

14. a. Label the reactive features, highlight the most reactive feature, and then highlight what it needs. Also, state if a carbocation, carbon radical, or carbanion will start to develop, and/or if aromatic character will be lost because of a reaction between these molecules. If a carbocation, carbon radical or carbanion starts to develop, label where that will occur.

methyl propanoate with xs NaBH₄ followed by acid

b. Use mechanism arrows to illustrate the reaction that occurs.

If applicable, use stabilization resources to deal with the carbocation, carbon radical, or carbanion that starts to develop during the reaction, and draw the structure of any resonance-stabilized intermediate.

Continue labelling and diagramming the reaction until you find the major stable product(s).

Finally, state the stereochemistry of the major product(s) and use either Fisher projection or perspective formula representations to illustrate that stereochemistry.

15. a. Label the reactive features, highlight the most reactive feature, and then highlight what it needs. Also, state if a carbocation, carbon radical, or carbanion will start to develop, and/or if aromatic character will be lost because of a reaction between these molecules. If a carbocation, carbon radical or carbanion starts to develop, label where that will occur.

p-propylaniline $+$ HNO$_3$ with tr. H$_2$SO$_4$

b. Use mechanism arrows to illustrate the reaction that occurs.

If applicable, use stabilization resources to deal with the carbocation, carbon radical, or carbanion that starts to develop during the reaction, and draw the structure of any resonance-stabilized intermediate.

Continue labelling and diagramming the reaction until you find the major stable product(s).

Finally, state the stereochemistry of the major product(s) and use either Fisher projection or perspective formula representations to illustrate that stereochemistry.

16. a. Label the reactive features, highlight the most reactive feature, and then highlight what it needs. Also, state if a carbocation, carbon radical, or carbanion will start to develop, and/or if aromatic character will be lost because of a reaction between these molecules. If a carbocation, carbon radical or carbanion starts to develop, label where that will occur.

$$H_3C\text{——}CH_2\text{——}CH_2\text{——}OH \quad \text{with} \quad TsCl \quad \text{and} \quad NaC\text{≡}N$$

b. Use mechanism arrows to illustrate the reaction that occurs.

If applicable, use stabilization resources to deal with the carbocation, carbon radical, or carbanion that starts to develop during the reaction, and draw the structure of any resonance-stabilized intermediate.

Continue labelling and diagramming the reaction until you find the major stable product(s).

Finally, state the stereochemistry of the major product(s) and use either Fisher projection or perspective formula representations to illustrate that stereochemistry.

17. a. Label the reactive features, highlight the most reactive feature, and then highlight what it needs. Also, state if a carbocation, carbon radical, or carbanion will start to develop, and/or if aromatic character will be lost because of a reaction between these molecules. If a carbocation, carbon radical or carbanion starts to develop, label where that will occur.

pent-1-yne + 1 equivalent of Br$_2$

b. Use mechanism arrows to illustrate the reaction that occurs.

If applicable, use stabilization resources to deal with the carbocation, carbon radical, or carbanion that starts to develop during the reaction, and draw the structure of any resonance-stabilized intermediate.

Continue labelling and diagramming the reaction until you find the major stable product(s).

Finally, state the stereochemistry of the major product(s) and use either Fisher projection or perspective formula representations to illustrate that stereochemistry.

18. a. Label the reactive features, highlight the most reactive feature, and then highlight what it needs. Also, state if a carbocation, carbon radical, or carbanion will start to develop, and/or if aromatic character will be lost because of a reaction between these molecules. If a carbocation, carbon radical or carbanion starts to develop, label where that will occur.

with tr. H_2SO_4

b. Use mechanism arrows to illustrate the reaction that occurs.

If applicable, use stabilization resources to deal with the carbocation, carbon radical, or carbanion that starts to develop during the reaction, and draw the structure of any resonance-stabilized intermediate.

Continue labelling and diagramming the reaction until you find the major stable product(s).

Finally, state the stereochemistry of the major product(s) and use either Fisher projection or perspective formula representations to illustrate that stereochemistry.

19. a. Label the reactive features, highlight the most reactive feature, and then highlight what it needs. Also, state if a carbocation, carbon radical, or carbanion will start to develop, and/or if aromatic character will be lost because of a reaction between these molecules. If a carbocation, carbon radical or carbanion starts to develop, label where that will occur.

(Z)-3-methylhept-3-ene + HBr and H_2O_2

b. Use mechanism arrows to illustrate the reaction that occurs.

If applicable, use stabilization resources to deal with the carbocation, carbon radical, or carbanion that starts to develop during the reaction, and draw the structure of any resonance-stabilized intermediate.

Continue labelling and diagramming the reaction until you find the major stable product(s).

Finally, state the stereochemistry of the major product(s) and use either Fisher projection or perspective formula representations to illustrate that stereochemistry.

20. a. Label the reactive features, highlight the most reactive feature, and then highlight what it needs. Also, state if a carbocation, carbon radical, or carbanion will start to develop, and/or if aromatic character will be lost because of a reaction between these molecules. If a carbocation, carbon radical or carbanion starts to develop, label where that will occur.

4-methylpenta-1,3-diene with the Z isomer of $CH_3CH=CHC\equiv N$

b. Use mechanism arrows to illustrate the reaction that occurs.

If applicable, use stabilization resources to deal with the carbocation, carbon radical, or carbanion that starts to develop during the reaction, and draw the structure of any resonance-stabilized intermediate.

Continue labelling and diagramming the reaction until you find the major stable product(s).

Finally, state the stereochemistry of the major product(s) and use either Fisher projection or perspective formula representations to illustrate that stereochemistry.

21. a. Label the reactive features, highlight the most reactive feature, and then highlight what it needs. Also, state if a carbocation, carbon radical, or carbanion will start to develop, and/or if aromatic character will be lost because of a reaction between these molecules. If a carbocation, carbon radical or carbanion starts to develop, label where that will occur.

fluorobenzene + 2-chloro-3-methyl butane + AlCl₃

b. Use mechanism arrows to illustrate the reaction that occurs.

If applicable, use stabilization resources to deal with the carbocation, carbon radical, or carbanion that starts to develop during the reaction, and draw the structure of any resonance-stabilized intermediate.

Continue labelling and diagramming the reaction until you find the major stable product(s).

Finally, state the stereochemistry of the major product(s) and use either Fisher projection or perspective formula representations to illustrate that stereochemistry.

22. a. Label the reactive features, highlight the most reactive feature, and then highlight what it needs. Also, state if a carbocation, carbon radical, or carbanion will start to develop, and/or if aromatic character will be lost because of a reaction between these molecules. If a carbocation, carbon radical or carbanion starts to develop, label where that will occur.

acetone in methanol with catalytic amounts of acid

b. Use mechanism arrows to illustrate the reaction that occurs.

If applicable, use stabilization resources to deal with the carbocation, carbon radical, or carbanion that starts to develop during the reaction, and draw the structure of any resonance-stabilized intermediate.

Continue labelling and diagramming the reaction until you find the major stable product(s).

Finally, state the stereochemistry of the major product(s) and use either Fisher projection or perspective formula representations to illustrate that stereochemistry.

23. a. Label the reactive features, highlight the most reactive feature, and then highlight what it needs. Also, state if a carbocation, carbon radical, or carbanion will start to develop, and/or if aromatic character will be lost because of a reaction between these molecules. If a carbocation, carbon radical or carbanion starts to develop, label where that will occur.

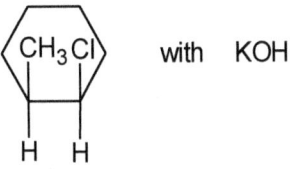 with   KOH

b. Use mechanism arrows to illustrate the reaction that occurs.

If applicable, use stabilization resources to deal with the carbocation, carbon radical, or carbanion that starts to develop during the reaction, and draw the structure of any resonance-stabilized intermediate.

Continue labelling and diagramming the reaction until you find the major stable product(s).

Finally, state the stereochemistry of the major product(s) and use either Fisher projection or perspective formula representations to illustrate that stereochemistry.

24. When an unknown substance is analyzed using mass spectrometry and infrared spectroscopy, the following data is obtained:

MS: M+ = 73, base peak = 44, and there is a major peak at an m/z of 58

IR:

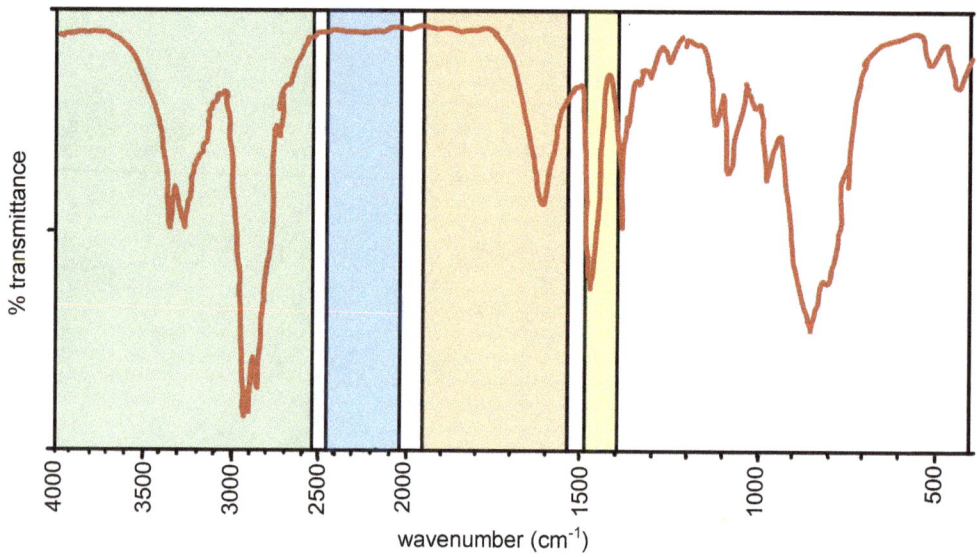

The compound is then reacted with cyclohexanone and catalytic amounts of acid. Draw the structure of the original substance, and then draw the product of the reaction.

Structure of the original compound: _____

Structure of the organic product: _____

Mash-up 3

1. (Chapters 1 and 2) Write the correct IUPAC name for the following:

Stereochemistry

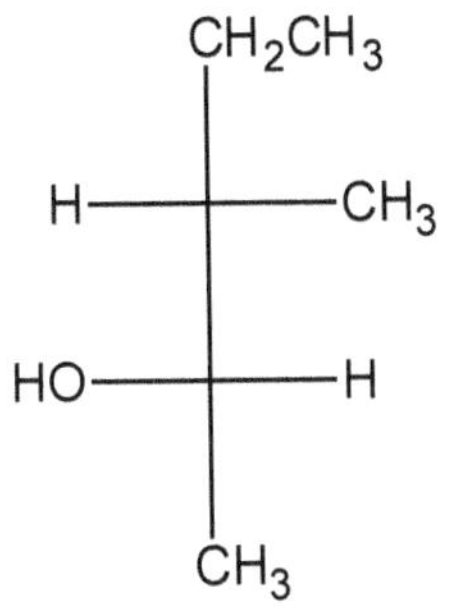

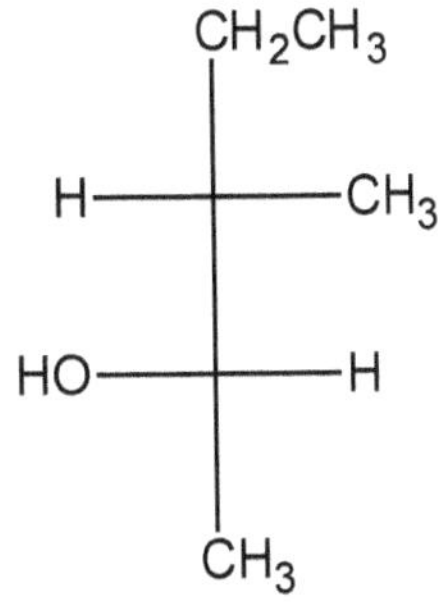

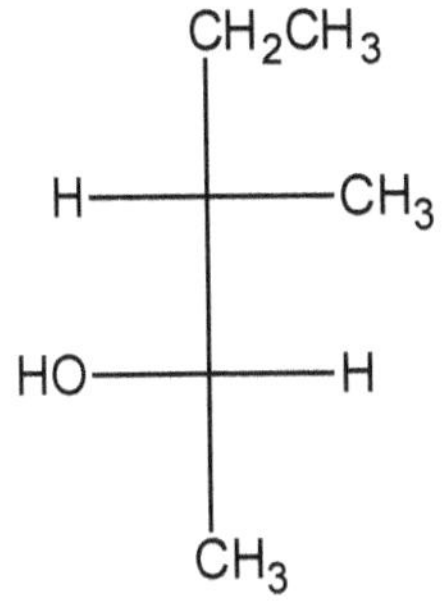

2. (Chapters 1 and 3) Write the correct IUPAC name and common name for the following:

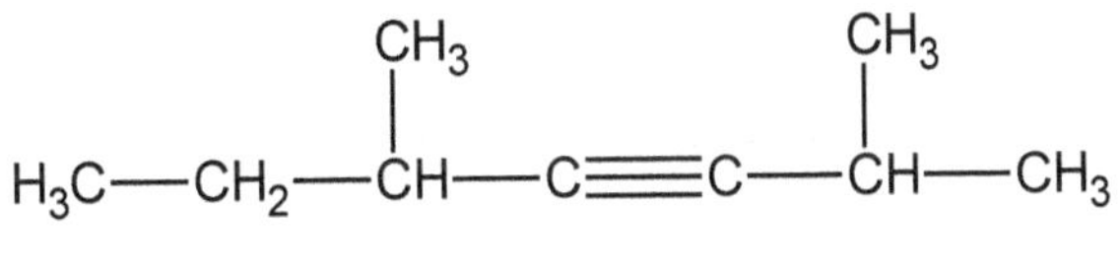

IUPAC Name: _____

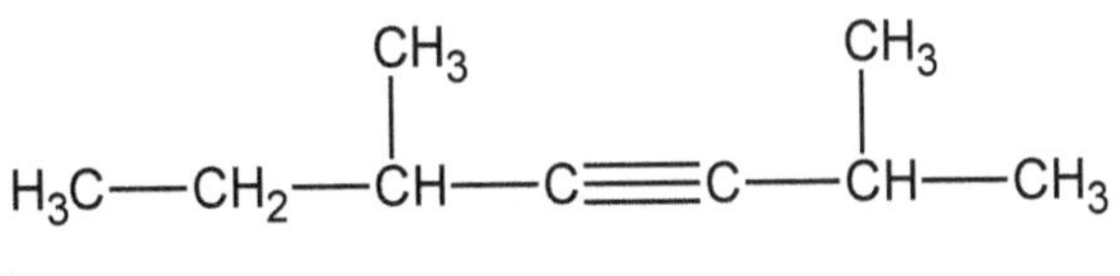

Common Name: _____

3. (Chapter 10) Write the correct common name for the following:

$$CH_3$$
$$HC-CH_3$$

(benzene ring with NH_2)

_____

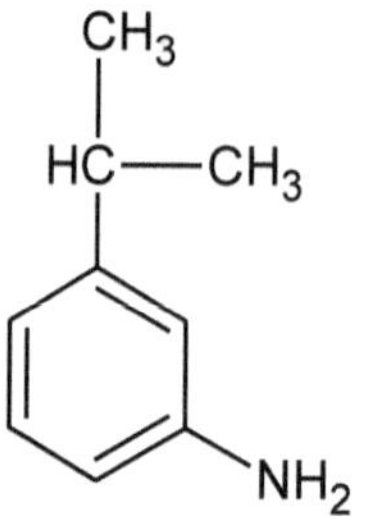

4. (Chapter 10) Write the correct common name for the following:

$$NO_2$$

(benzene ring with Cl and OH)

_____

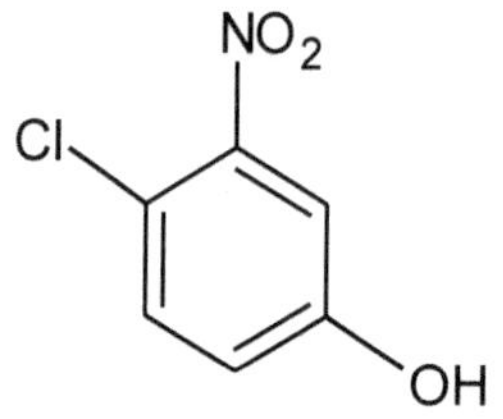

5. (Chapter 11) Write the correct IUPAC name and common name for the following:

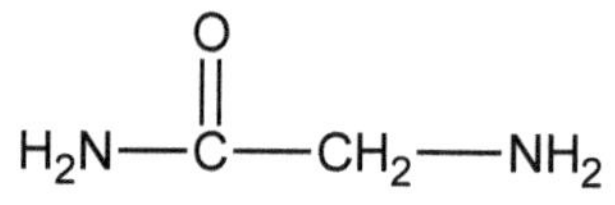

$$H_2N-\overset{\overset{O}{\|}}{C}-CH_2-NH_2$$

IUPAC Name: _____

$$H_2N-\overset{\overset{O}{\|}}{C}-CH_2-NH_2$$

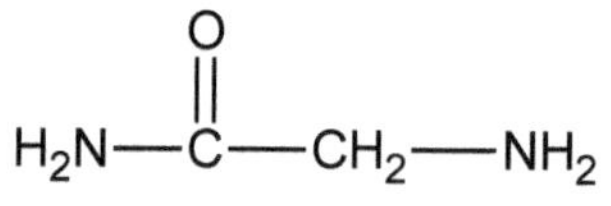

Common Name _____

6. (Chapter 11) Write the correct IUPAC name for the following:

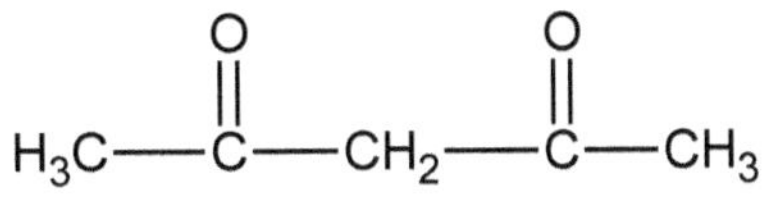

$$H_3C-\overset{\overset{\displaystyle O}{\|}}{C}-CH_2-\overset{\overset{\displaystyle O}{\|}}{C}-CH_3$$

_____

7. (Chapter 13) Write two correct IUPAC names and a common name for the following:

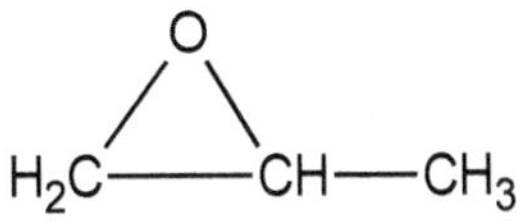

$$H_2C \underset{\diagdown \diagup}{\overset{O}{\triangle}} CH-CH_3$$

IUPAC Name: _____

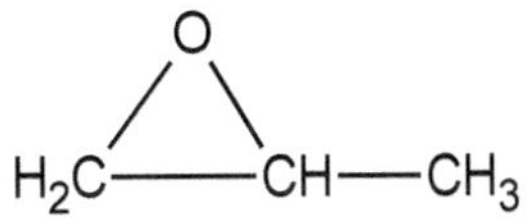

$$H_2C \underset{\diagdown \diagup}{\overset{O}{\triangle}} CH-CH_3$$

IUPAC Name: _____

$$H_2C \underset{\diagdown \diagup}{\overset{O}{\triangle}} CH-CH_3$$

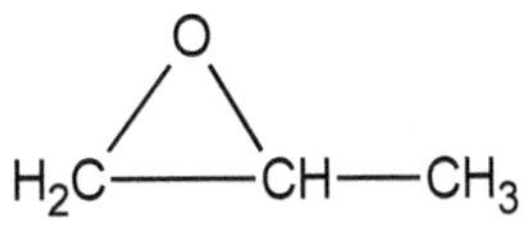

Common Name _____

8. (Chapters 5 and 6) What type of instability will need to be addressed in the intermediate that is created when a radical reacts with bond electrons?

_____

9. (Chapter 7) What type of instability will need to be addressed in the intermediate that is created when a pi bond reacts with an electron-deficient hydrogen?

_____

10. (Chapters 6, 7, and 8) What general methods could potentially be used to stabilize a carbocation?

_____

_____

_____

_____

_____

_____

(Chapters 5–13) For each of the following,

11. a. Label the reactive features, highlight the most reactive feature, and then highlight what it needs. Also, state if a carbocation, carbon radical, or carbanion will start to develop, and/or if aromatic character will be lost because of a reaction between these molecules. If a carbocation, carbon radical or carbanion starts to develop, label where that will occur.

2-methylcyclopenta-1,3-diene + HCl

b. Use mechanism arrows to illustrate the reaction that occurs.

If applicable, use stabilization resources to deal with the carbocation, carbon radical, or carbanion that starts to develop during the reaction, and draw the structure of any resonance-stabilized intermediate.

Continue labelling and diagramming the reaction until you find the major stable product(s).

Finally, state the stereochemistry of the major product(s) and use either Fisher projection or perspective formula representations to illustrate that stereochemistry.

12. a. Label the reactive features, highlight the most reactive feature, and then highlight what it needs. Also, state if a carbocation, carbon radical, or carbanion will start to develop, and/or if aromatic character will be lost because of a reaction between these molecules. If a carbocation, carbon radical or carbanion starts to develop, label where that will occur.

HN——CH₃

CH₂——CH₂——CH₃   +   2-chloro-3-methyl butane   +   AlCl₃

b. Use mechanism arrows to illustrate the reaction that occurs.

If applicable, use stabilization resources to deal with the carbocation, carbon radical, or carbanion that starts to develop during the reaction, and draw the structure of any resonance-stabilized intermediate.

Continue labelling and diagramming the reaction until you find the major stable product(s).

Finally, state the stereochemistry of the major product(s) and use either Fisher projection or perspective formula representations to illustrate that stereochemistry.

13. a. Label the reactive features, highlight the most reactive feature, and then highlight what it needs. Also, state if a carbocation, carbon radical, or carbanion will start to develop, and/or if aromatic character will be lost because of a reaction between these molecules. If a carbocation, carbon radical or carbanion starts to develop, label where that will occur.

   butyric acid + sodium bromide

   b. Use mechanism arrows to illustrate the reaction that occurs.

   If applicable, use stabilization resources to deal with the carbocation, carbon radical, or carbanion that starts to develop during the reaction, and draw the structure of any resonance-stabilized intermediate.

   Continue labelling and diagramming the reaction until you find the major stable product(s).

   Finally, state the stereochemistry of the major product(s) and use either Fisher projection or perspective formula representations to illustrate that stereochemistry.

14. a. Label the reactive features, highlight the most reactive feature, and then highlight what it needs. Also, state if a carbocation, carbon radical, or carbanion will start to develop, and/or if aromatic character will be lost because of a reaction between these molecules. If a carbocation, carbon radical or carbanion starts to develop, label where that will occur.

3,3-dimethylcyclohexene + Hg(OAc)$_2$ in water, followed by NaBH$_4$

b. Use mechanism arrows to illustrate the reaction that occurs.

If applicable, use stabilization resources to deal with the carbocation, carbon radical, or carbanion that starts to develop during the reaction, and draw the structure of any resonance-stabilized intermediate.

Continue labelling and diagramming the reaction until you find the major stable product(s).

Finally, state the stereochemistry of the major product(s) and use either Fisher projection or perspective formula representations to illustrate that stereochemistry.

15. a. Label the reactive features, highlight the most reactive feature, and then highlight what it needs. Also, state if a carbocation, carbon radical, or carbanion will start to develop, and/or if aromatic character will be lost because of a reaction between these molecules. If a carbocation, carbon radical or carbanion starts to develop, label where that will occur.

bromoethane with potassium methoxide in methanol

b. Use mechanism arrows to illustrate the reaction that occurs.

If applicable, use stabilization resources to deal with the carbocation, carbon radical, or carbanion that starts to develop during the reaction, and draw the structure of any resonance-stabilized intermediate.

Continue labelling and diagramming the reaction until you find the major stable product(s).

Finally, state the stereochemistry of the major product(s) and use either Fisher projection or perspective formula representations to illustrate that stereochemistry.

16. a. Label the reactive features, highlight the most reactive feature, and then highlight what it needs. Also, state if a carbocation, carbon radical, or carbanion will start to develop, and/or if aromatic character will be lost because of a reaction between these molecules. If a carbocation, carbon radical or carbanion starts to develop, label where that will occur.

1-methylcyclopentene in ethanol with a tr. of acid

b. Use mechanism arrows to illustrate the reaction that occurs.

If applicable, use stabilization resources to deal with the carbocation, carbon radical, or carbanion that starts to develop during the reaction, and draw the structure of any resonance-stabilized intermediate.

Continue labelling and diagramming the reaction until you find the major stable product(s).

Finally, state the stereochemistry of the major product(s) and use either Fisher projection or perspective formula representations to illustrate that stereochemistry.

17. a. Label the reactive features, highlight the most reactive feature, and then highlight what it needs. Also, state if a carbocation, carbon radical, or carbanion will start to develop, and/or if aromatic character will be lost because of a reaction between these molecules. If a carbocation, carbon radical or carbanion starts to develop, label where that will occur.

$$\text{m-isopropylbenzaldehyde} \ + \text{H}_2\text{SO}_4$$

b. Use mechanism arrows to illustrate the reaction that occurs.

If applicable, use stabilization resources to deal with the carbocation, carbon radical, or carbanion that starts to develop during the reaction, and draw the structure of any resonance-stabilized intermediate.

Continue labelling and diagramming the reaction until you find the major stable product(s).

Finally, state the stereochemistry of the major product(s) and use either Fisher projection or perspective formula representations to illustrate that stereochemistry.

18. a. Label the reactive features, highlight the most reactive feature, and then highlight what it needs. Also, state if a carbocation, carbon radical, or carbanion will start to develop, and/or if aromatic character will be lost because of a reaction between these molecules. If a carbocation, carbon radical or carbanion starts to develop, label where that will occur.

(S)-2,3-dimethylpentan-3-ol with concentrated hydrochloric acid

b. Use mechanism arrows to illustrate the reaction that occurs.

If applicable, use stabilization resources to deal with the carbocation, carbon radical, or carbanion that starts to develop during the reaction, and draw the structure of any resonance-stabilized intermediate.

Continue labelling and diagramming the reaction until you find the major stable product(s).

Finally, state the stereochemistry of the major product(s) and use either Fisher projection or perspective formula representations to illustrate that stereochemistry.

19. a. Label the reactive features, highlight the most reactive feature, and then highlight what it needs. Also, state if a carbocation, carbon radical, or carbanion will start to develop, and/or if aromatic character will be lost because of a reaction between these molecules. If a carbocation, carbon radical or carbanion starts to develop, label where that will occur.

pentan-3-one + sodium hydroxide

b. Use mechanism arrows to illustrate the reaction that occurs.

If applicable, use stabilization resources to deal with the carbocation, carbon radical, or carbanion that starts to develop during the reaction, and draw the structure of any resonance-stabilized intermediate.

Continue labelling and diagramming the reaction until you find the major stable product(s).

Finally, state the stereochemistry of the major product(s) and use either Fisher projection or perspective formula representations to illustrate that stereochemistry.

20. a. Label the reactive features, highlight the most reactive feature, and then highlight what it needs. Also, state if a carbocation, carbon radical, or carbanion will start to develop, and/or if aromatic character will be lost because of a reaction between these molecules. If a carbocation, carbon radical or carbanion starts to develop, label where that will occur.

toluene + Cl_2 and $FeCl_3$

b. Use mechanism arrows to illustrate the reaction that occurs.

If applicable, use stabilization resources to deal with the carbocation, carbon radical, or carbanion that starts to develop during the reaction, and draw the structure of any resonance-stabilized intermediate.

Continue labelling and diagramming the reaction until you find the major stable product(s).

Finally, state the stereochemistry of the major product(s) and use either Fisher projection or perspective formula representations to illustrate that stereochemistry.

21. When an unknown substance is analyzed using mass spectrometry and infrared spectroscopy, it gives the following data:

MS: The M+ and base peak have an m/z value of 54

IR:

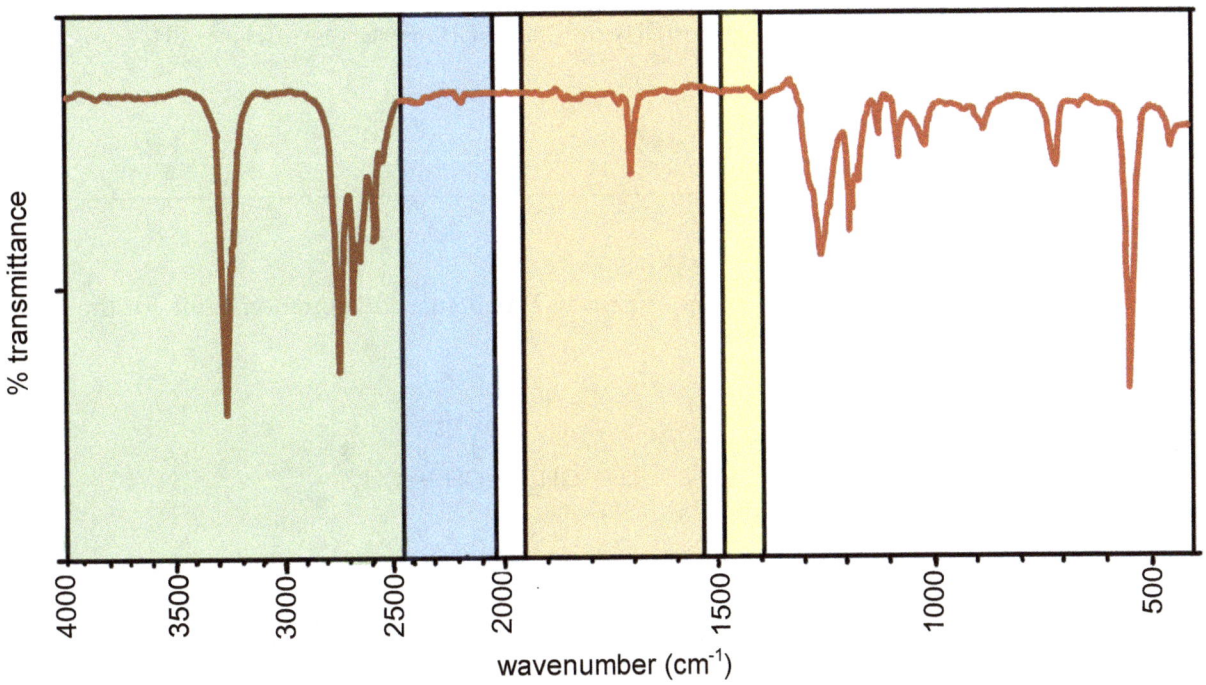

wavenumber (cm⁻¹)

The substance is then reacted with HgSO₄ and H₂SO₄ in water. Draw the structure of the unknown reactant, and then draw the organic product that results from the reaction.

Structure of the original compound: _____

Structure of the organic product: _____

Mash-up 4

1. (Chapter 1) Write the correct IUPAC name for the following:

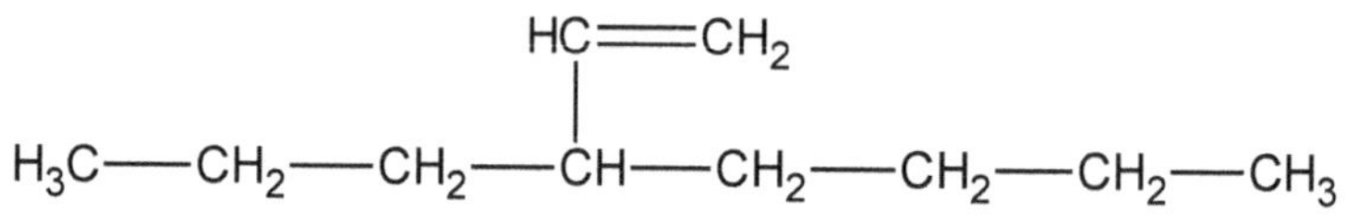

$$HC\!\!=\!\!CH_2$$
$$H_3C\!-\!CH_2\!-\!CH_2\!-\!CH\!-\!CH_2\!-\!CH_2\!-\!CH_2\!-\!CH_3$$

_____

2. (Chapters 1 and 3) Write the correct IUPAC name and common name for the following:

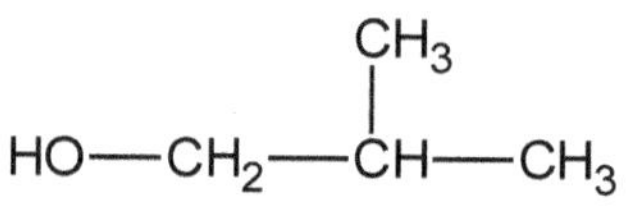

$$CH_3$$
$$HO\!-\!CH_2\!-\!CH\!-\!CH_3$$

IUPAC Name: _____

$$CH_3$$
$$HO\!-\!CH_2\!-\!CH\!-\!CH_3$$

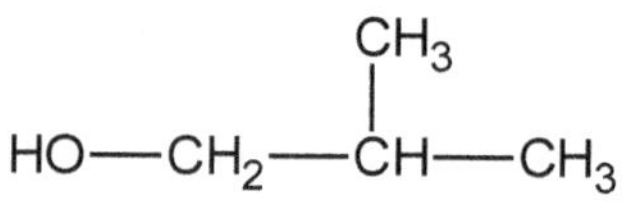

Common Name _____

3. (Chapter 10) Write the correct common name for the following:

$$CH_3$$
$$NH_2$$

_____

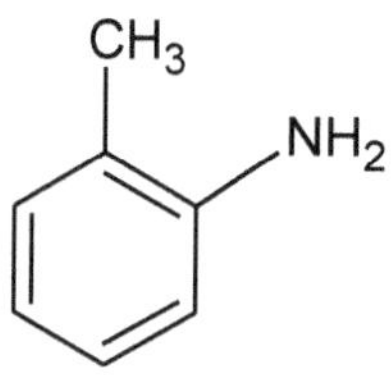

4. (Chapter 10) Write the correct common name for the following:

_____

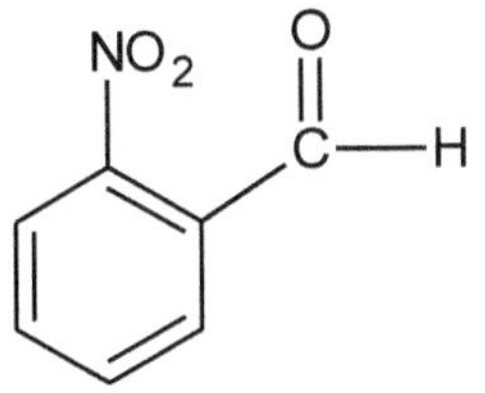

5. (Chapter 11) Write the correct IUPAC name and derived name for the following:

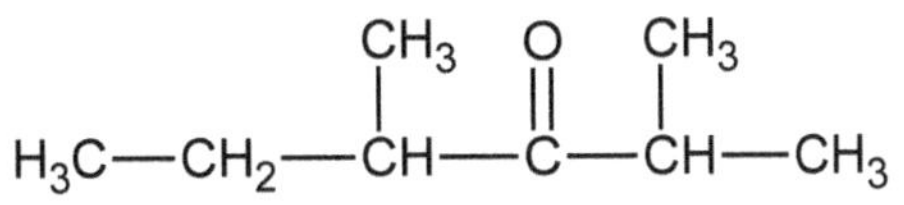

IUPAC Name: _____

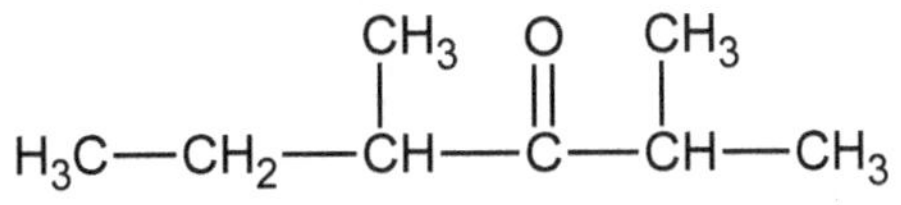

Derived Name _____

6. (Chapter 11) Write the correct common name for the following:

_____

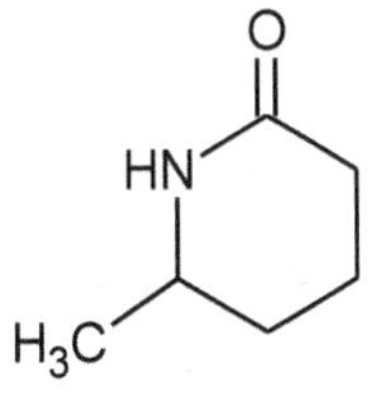

7. (Chapter 13) Write two correct IUPAC names for the following:

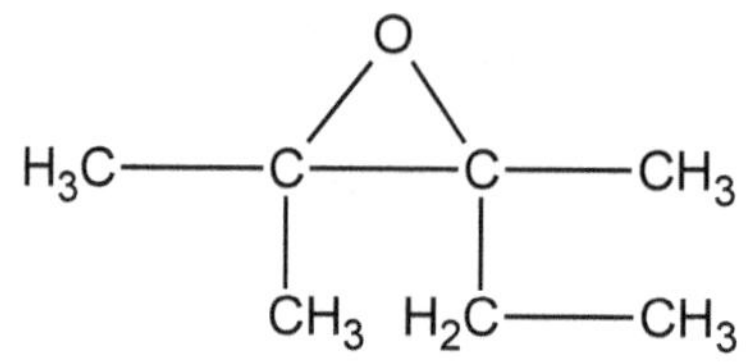

IUPAC Name: _____

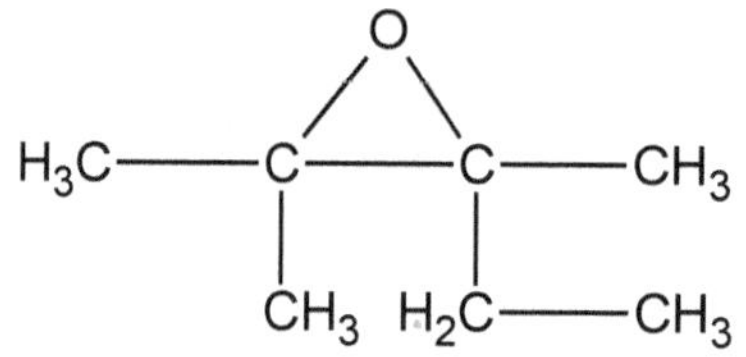

IUPAC Name: _____

8. (Chapter 6) What type of instability will need to be addressed in the intermediate that is created when a carbon that has a good leaving group is in a protic polar solvent?

_____

9. (Chapter 10) What type of instability will need to be addressed in the intermediate that is created when an aromatic compound reacts with a very reactive electrophile?

_____

10. (Chapter 13) What general resources could potentially be used to react with a carbon that has a partial positive charge?

_____

_____

_____

(Chapters 5–13) For each of the following,

11. a. Label the reactive features, highlight the most reactive feature, and then highlight what it needs. Also, state if a carbocation, carbon radical, or carbanion will start to develop, and/or if aromatic character will be lost because of a reaction between these molecules. If a carbocation, carbon radical or carbanion starts to develop, label where that will occur.

$$\begin{array}{c} \quad\ \text{CH}_3 \quad\ \text{CH}_3 \\ \quad\ | \qquad\ | \\ \text{H}_3\text{C—CH}_2\!\!\underset{\text{R}}{\text{—}}\!\text{C}\!\text{—}\!\!\underset{\text{R}}{\text{CH}}\!\text{—CH}_2\text{—CH}_3 \qquad \text{with} \quad \text{NaOH} \\ \quad\ | \\ \quad\ \text{Cl} \end{array}$$

b. Use mechanism arrows to illustrate the reaction that occurs.

If applicable, use stabilization resources to deal with the carbocation, carbon radical, or carbanion that starts to develop during the reaction, and draw the structure of any resonance-stabilized intermediate.

Continue labelling and diagramming the reaction until you find the major stable product(s).

Finally, state the stereochemistry of the major product(s) and use either Fisher projection or perspective formula representations to illustrate that stereochemistry.

12. a. Label the reactive features, highlight the most reactive feature, and then highlight what it needs. Also, state if a carbocation, carbon radical, or carbanion will start to develop, and/or if aromatic character will be lost because of a reaction between these molecules. If a carbocation, carbon radical or carbanion starts to develop, label where that will occur.

pent-3-en-2-one with potassium ethoxide

b. Use mechanism arrows to illustrate the reaction that occurs.

If applicable, use stabilization resources to deal with the carbocation, carbon radical, or carbanion that starts to develop during the reaction, and draw the structure of any resonance-stabilized intermediate.

Continue labelling and diagramming the reaction until you find the major stable product(s).

Finally, state the stereochemistry of the major product(s) and use either Fisher projection or perspective formula representations to illustrate that stereochemistry.

13. a. Label the reactive features, highlight the most reactive feature, and then highlight what it needs. Also, state if a carbocation, carbon radical, or carbanion will start to develop, and/or if aromatic character will be lost because of a reaction between these molecules. If a carbocation, carbon radical or carbanion starts to develop, label where that will occur.

3-methylcyclohexene + BH_3 followed by H_2O_2, ^-OH, and H_2O

b. Use mechanism arrows to illustrate the reaction that occurs.

If applicable, use stabilization resources to deal with the carbocation, carbon radical, or carbanion that starts to develop during the reaction, and draw the structure of any resonance-stabilized intermediate.

Continue labelling and diagramming the reaction until you find the major stable product(s).

Finally, state the stereochemistry of the major product(s) and use either Fisher projection or perspective formula representations to illustrate that stereochemistry.

14. a. Label the reactive features, highlight the most reactive feature, and then highlight what it needs. Also, state if a carbocation, carbon radical, or carbanion will start to develop, and/or if aromatic character will be lost because of a reaction between these molecules. If a carbocation, carbon radical or carbanion starts to develop, label where that will occur.

$$+ H_2SO_4$$

b. Use mechanism arrows to illustrate the reaction that occurs.

If applicable, use stabilization resources to deal with the carbocation, carbon radical, or carbanion that starts to develop during the reaction, and draw the structure of any resonance-stabilized intermediate.

Continue labelling and diagramming the reaction until you find the major stable product(s).

Finally, state the stereochemistry of the major product(s) and use either Fisher projection or perspective formula representations to illustrate that stereochemistry.

15. a. Label the reactive features, highlight the most reactive feature, and then highlight what it needs. Also, state if a carbocation, carbon radical, or carbanion will start to develop, and/or if aromatic character will be lost because of a reaction between these molecules. If a carbocation, carbon radical or carbanion starts to develop, label where that will occur.

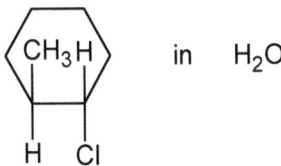 in H$_2$O

b. Use mechanism arrows to illustrate the reaction that occurs.

If applicable, use stabilization resources to deal with the carbocation, carbon radical, or carbanion that starts to develop during the reaction, and draw the structure of any resonance-stabilized intermediate.

Continue labelling and diagramming the reaction until you find the major stable product(s).

Finally, state the stereochemistry of the major product(s) and use either Fisher projection or perspective formula representations to illustrate that stereochemistry.

16. a. Label the reactive features, highlight the most reactive feature, and then highlight what it needs. Also, state if a carbocation, carbon radical, or carbanion will start to develop, and/or if aromatic character will be lost because of a reaction between these molecules. If a carbocation, carbon radical or carbanion starts to develop, label where that will occur.

(R)-3-methylpent-1-ene + Br$_2$ in ethanol

b. Use mechanism arrows to illustrate the reaction that occurs.

If applicable, use stabilization resources to deal with the carbocation, carbon radical, or carbanion that starts to develop during the reaction, and draw the structure of any resonance-stabilized intermediate.

Continue labelling and diagramming the reaction until you find the major stable product(s).

Finally, state the stereochemistry of the major product(s) and use either Fisher projection or perspective formula representations to illustrate that stereochemistry.

17. a. Label the reactive features, highlight the most reactive feature, and then highlight what it needs. Also, state if a carbocation, carbon radical, or carbanion will start to develop, and/or if aromatic character will be lost because of a reaction between these molecules. If a carbocation, carbon radical or carbanion starts to develop, label where that will occur.

m-propylphenol + Cl$_2$ and FeCl$_3$

b. Use mechanism arrows to illustrate the reaction that occurs.

If applicable, use stabilization resources to deal with the carbocation, carbon radical, or carbanion that starts to develop during the reaction, and draw the structure of any resonance-stabilized intermediate.

Continue labelling and diagramming the reaction until you find the major stable product(s).

Finally, state the stereochemistry of the major product(s) and use either Fisher projection or perspective formula representations to illustrate that stereochemistry.

18. a. Label the reactive features, highlight the most reactive feature, and then highlight what it needs. Also, state if a carbocation, carbon radical, or carbanion will start to develop, and/or if aromatic character will be lost because of a reaction between these molecules. If a carbocation, carbon radical or carbanion starts to develop, label where that will occur.

       3,4-dimethylpenta-1,3-diene with HBr

b. Use mechanism arrows to illustrate the reaction that occurs.

If applicable, use stabilization resources to deal with the carbocation, carbon radical, or carbanion that starts to develop during the reaction, and draw the structure of any resonance-stabilized intermediate.

Continue labelling and diagramming the reaction until you find the major stable product(s).

Finally, state the stereochemistry of the major product(s) and use either Fisher projection or perspective formula representations to illustrate that stereochemistry.

19. a. Label the reactive features, highlight the most reactive feature, and then highlight what it needs. Also, state if a carbocation, carbon radical, or carbanion will start to develop, and/or if aromatic character will be lost because of a reaction between these molecules. If a carbocation, carbon radical or carbanion starts to develop, label where that will occur.

valeryl bromide + potassium methoxide

b. Use mechanism arrows to illustrate the reaction that occurs.

If applicable, use stabilization resources to deal with the carbocation, carbon radical, or carbanion that starts to develop during the reaction, and draw the structure of any resonance-stabilized intermediate.

Continue labelling and diagramming the reaction until you find the major stable product(s).

Finally, state the stereochemistry of the major product(s) and use either Fisher projection or perspective formula representations to illustrate that stereochemistry.

20. a. Label the reactive features, highlight the most reactive feature, and then highlight what it needs. Also, state if a carbocation, carbon radical, or carbanion will start to develop, and/or if aromatic character will be lost because of a reaction between these molecules. If a carbocation, carbon radical or carbanion starts to develop, label where that will occur.

nitrobenzene + H_2SO_4

b. Use mechanism arrows to illustrate the reaction that occurs.

If applicable, use stabilization resources to deal with the carbocation, carbon radical, or carbanion that starts to develop during the reaction, and draw the structure of any resonance-stabilized intermediate.

Continue labelling and diagramming the reaction until you find the major stable product(s).

Finally, state the stereochemistry of the major product(s) and use either Fisher projection or perspective formula representations to illustrate that stereochemistry.

21. When an unknown compound is analyzed using mass spectrometry and infrared spectroscopy, the following data is obtained:

MS: M+ = 88 (missing), significant peaks at m/z = 70, and m/z = 31.

IR:

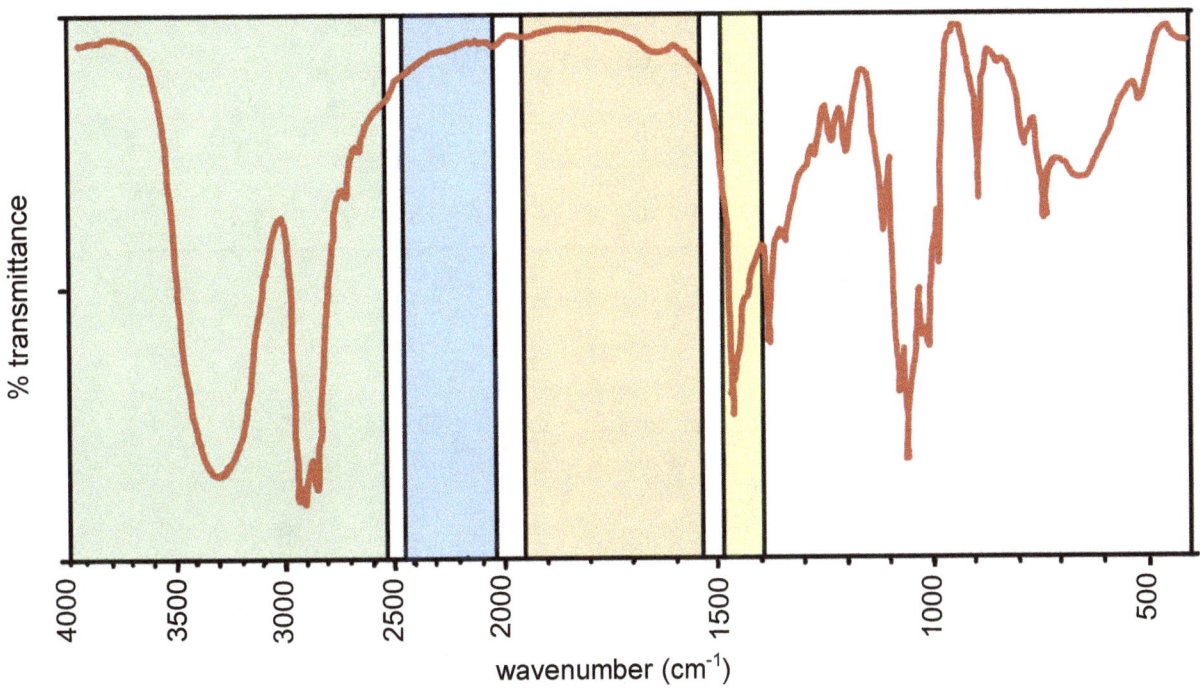

Draw the structure of the molecule, and then propose a strategy for synthesizing the compound from an aldehyde or ketone.

Structure of the original compound: _____

Synthesis strategy:

Mash-up 5

1. (Chapters 1 and 2) Assign an IUPAC name to the following:

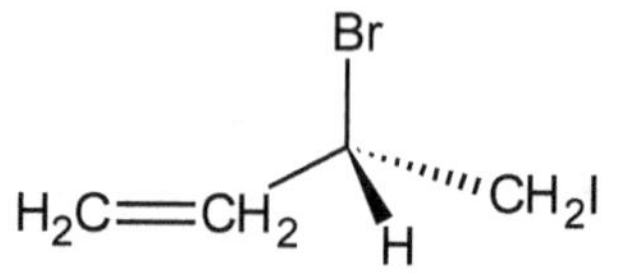

_____

2. (Chapter 10) Write the correct common name for the following:

_____

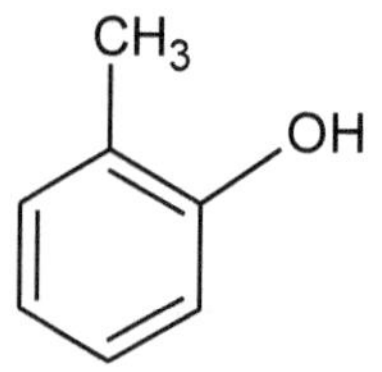

3. (Chapter 10) Write the correct common name for the following:

_____

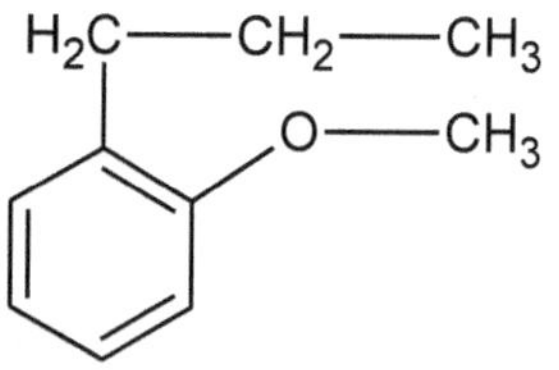

4. (Chapter 11) Write the correct IUPAC name, common name, and derived name for the following:

$$H_3C—C≡N$$

IUPAC Name: _____

$$H_3C—C≡N$$

Common Name _____

$$H_3C—C≡N$$

Derived Name _____

5. (Chapter 11) Write the correct common name for the following:

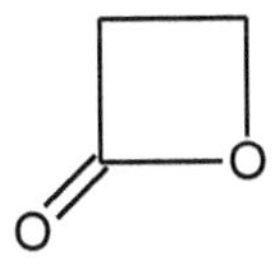

_____

6. (Chapter 13) Write two correct IUPAC names for the following:

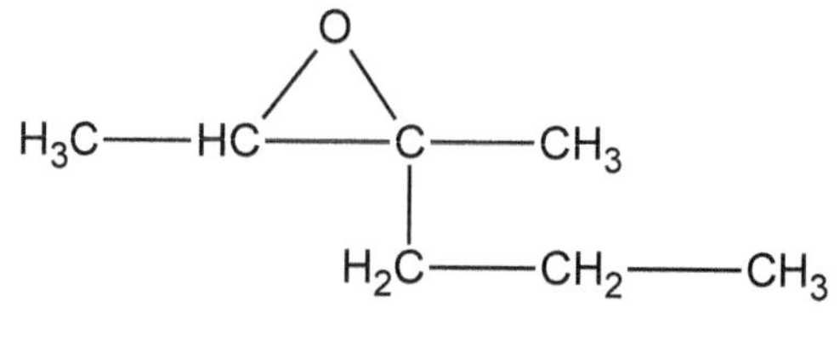

IUPAC Name 1: _____

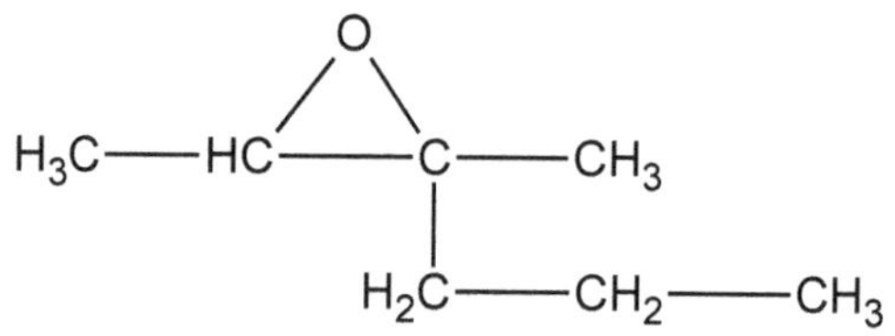

IUPAC Name 2: _____

7. (Chapter 12) What two possibilities exist for reacting a carbonyl group that doesn't have a conjugated pi bond, why do these two possibilities exist, and under what conditions would you select each?

1.) _____

_____

2.) _____

_____

(Chapters 5–13) For each of the following,

8. a. Label the reactive features, highlight the most reactive feature, and then highlight what it needs. Also, state if a carbocation, carbon radical, or carbanion will start to develop, and/or if aromatic character will be lost because of a reaction between these molecules. If a carbocation, carbon radical or carbanion starts to develop, label where that will occur.

(Z)-4,5-dimethyloct-4-ene + Hg(OAc)$_2$ in water, followed by NaBH$_4$

b. Use mechanism arrows to illustrate the reaction that occurs.

If applicable, use stabilization resources to deal with the carbocation, carbon radical, or carbanion that starts to develop during the reaction, and draw the structure of any resonance-stabilized intermediate.

Continue labelling and diagramming the reaction until you find the major stable product(s).

Finally, state the stereochemistry of the major product(s) and use either Fisher projection or perspective formula representations to illustrate that stereochemistry.

9. a. Label the reactive features, highlight the most reactive feature, and then highlight what it needs. Also, state if a carbocation, carbon radical, or carbanion will start to develop, and/or if aromatic character will be lost because of a reaction between these molecules. If a carbocation, carbon radical or carbanion starts to develop, label where that will occur.

acetone + LiAlH₄ followed by acid

b. Use mechanism arrows to illustrate the reaction that occurs.

If applicable, use stabilization resources to deal with the carbocation, carbon radical, or carbanion that starts to develop during the reaction, and draw the structure of any resonance-stabilized intermediate.

Continue labelling and diagramming the reaction until you find the major stable product(s).

Finally, state the stereochemistry of the major product(s) and use either Fisher projection or perspective formula representations to illustrate that stereochemistry.

10. a. Label the reactive features, highlight the most reactive feature, and then highlight what it needs. Also, state if a carbocation, carbon radical, or carbanion will start to develop, and/or if aromatic character will be lost because of a reaction between these molecules. If a carbocation, carbon radical or carbanion starts to develop, label where that will occur.

benzenesulfonic acid + 2-methylpropanoyl chloride + AlCl$_3$

b. Use mechanism arrows to illustrate the reaction that occurs.

If applicable, use stabilization resources to deal with the carbocation, carbon radical, or carbanion that starts to develop during the reaction, and draw the structure of any resonance-stabilized intermediate.

Continue labelling and diagramming the reaction until you find the major stable product(s).

Finally, state the stereochemistry of the major product(s) and use either Fisher projection or perspective formula representations to illustrate that stereochemistry.

11. a. Label the reactive features, highlight the most reactive feature, and then highlight what it needs. Also, state if a carbocation, carbon radical, or carbanion will start to develop, and/or if aromatic character will be lost because of a reaction between these molecules. If a carbocation, carbon radical or carbanion starts to develop, label where that will occur.

(2R,3S)-2-chloro-3-methylpentane with potassium ethoxide in ethanol 8

b. Use mechanism arrows to illustrate the reaction that occurs.

If applicable, use stabilization resources to deal with the carbocation, carbon radical, or carbanion that starts to develop during the reaction, and draw the structure of any resonance-stabilized intermediate.

Continue labelling and diagramming the reaction until you find the major stable product(s).

Finally, state the stereochemistry of the major product(s) and use either Fisher projection or perspective formula representations to illustrate that stereochemistry.

12. a. Label the reactive features, highlight the most reactive feature, and then highlight what it needs. Also, state if a carbocation, carbon radical, or carbanion will start to develop, and/or if aromatic character will be lost because of a reaction between these molecules. If a carbocation, carbon radical or carbanion starts to develop, label where that will occur.

1,2-dimethylcyclopentene + HBr

b. Use mechanism arrows to illustrate the reaction that occurs.

If applicable, use stabilization resources to deal with the carbocation, carbon radical, or carbanion that starts to develop during the reaction, and draw the structure of any resonance-stabilized intermediate.

Continue labelling and diagramming the reaction until you find the major stable product(s).

Finally, state the stereochemistry of the major product(s) and use either Fisher projection or perspective formula representations to illustrate that stereochemistry.

13. a. Label the reactive features, highlight the most reactive feature, and then highlight what it needs. Also, state if a carbocation, carbon radical, or carbanion will start to develop, and/or if aromatic character will be lost because of a reaction between these molecules. If a carbocation, carbon radical or carbanion starts to develop, label where that will occur.

pentan-3-en-2-one + CH_3MgBr followed by acid

b. Use mechanism arrows to illustrate the reaction that occurs.

If applicable, use stabilization resources to deal with the carbocation, carbon radical, or carbanion that starts to develop during the reaction, and draw the structure of any resonance-stabilized intermediate.

Continue labelling and diagramming the reaction until you find the major stable product(s).

Finally, state the stereochemistry of the major product(s) and use either Fisher projection or perspective formula representations to illustrate that stereochemistry.

14. a. Label the reactive features, highlight the most reactive feature, and then highlight what it needs. Also, state if a carbocation, carbon radical, or carbanion will start to develop, and/or if aromatic character will be lost because of a reaction between these molecules. If a carbocation, carbon radical or carbanion starts to develop, label where that will occur.

phenol + 2-methylpropanoyl chloride + AlCl$_3$

b. Use mechanism arrows to illustrate the reaction that occurs.

If applicable, use stabilization resources to deal with the carbocation, carbon radical, or carbanion that starts to develop during the reaction, and draw the structure of any resonance-stabilized intermediate.

Continue labelling and diagramming the reaction until you find the major stable product(s).

Finally, state the stereochemistry of the major product(s) and use either Fisher projection or perspective formula representations to illustrate that stereochemistry.

15. a. Label the reactive features, highlight the most reactive feature, and then highlight what it needs. Also, state if a carbocation, carbon radical, or carbanion will start to develop, and/or if aromatic character will be lost because of a reaction between these molecules. If a carbocation, carbon radical or carbanion starts to develop, label where that will occur.

(R)-butan-2-ol with concentrated H_3PO_4

b. Use mechanism arrows to illustrate the reaction that occurs.

If applicable, use stabilization resources to deal with the carbocation, carbon radical, or carbanion that starts to develop during the reaction, and draw the structure of any resonance-stabilized intermediate.

Continue labelling and diagramming the reaction until you find the major stable product(s).

Finally, state the stereochemistry of the major product(s) and use either Fisher projection or perspective formula representations to illustrate that stereochemistry.

16. a. Label the reactive features, highlight the most reactive feature, and then highlight what it needs. Also, state if a carbocation, carbon radical, or carbanion will start to develop, and/or if aromatic character will be lost because of a reaction between these molecules. If a carbocation, carbon radical or carbanion starts to develop, label where that will occur.

1-methylcyclopenta-1,3-diene with HCl

b. Use mechanism arrows to illustrate the reaction that occurs.

If applicable, use stabilization resources to deal with the carbocation, carbon radical, or carbanion that starts to develop during the reaction, and draw the structure of any resonance-stabilized intermediate.

Continue labelling and diagramming the reaction until you find the major stable product(s).

Finally, state the stereochemistry of the major product(s) and use either Fisher projection or perspective formula representations to illustrate that stereochemistry.

17. a. Label the reactive features, highlight the most reactive feature, and then highlight what it needs. Also, state if a carbocation, carbon radical, or carbanion will start to develop, and/or if aromatic character will be lost because of a reaction between these molecules. If a carbocation, carbon radical or carbanion starts to develop, label where that will occur.

p-propylphenol + H_2SO_4

b. Use mechanism arrows to illustrate the reaction that occurs.

If applicable, use stabilization resources to deal with the carbocation, carbon radical, or carbanion that starts to develop during the reaction, and draw the structure of any resonance-stabilized intermediate.

Continue labelling and diagramming the reaction until you find the major stable product(s).

Finally, state the stereochemistry of the major product(s) and use either Fisher projection or perspective formula representations to illustrate that stereochemistry.

18. Determine the structure of the compound represented by the following MS and IR spectra, then write the product that is formed when that molecule reacts with CH_3MgBr.

MS:

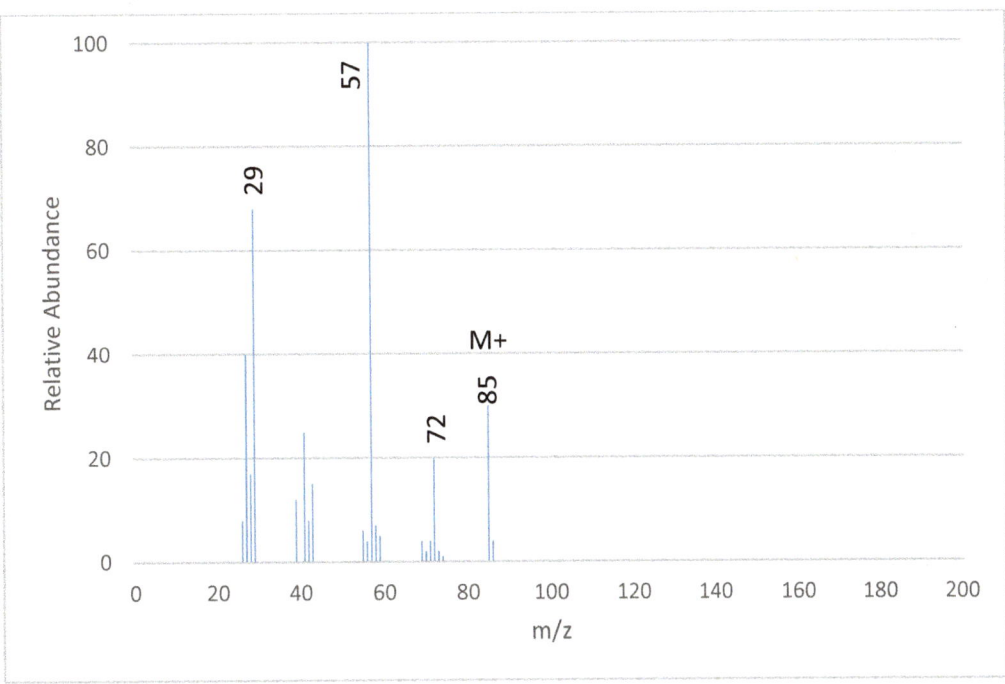

IR:

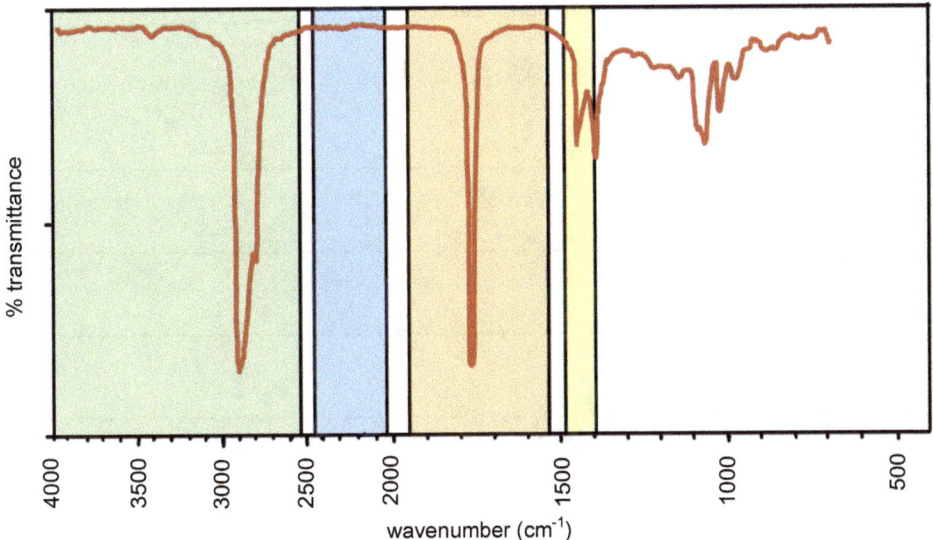

Structure of the original compound: _____

Structure of the organic product: _____

Summary of Concepts and Analysis Methods

Summary of the Chapter:

_____

_____

_____

_____

_____

_____

_____

Questions You Should Ask In Class:

_____

_____

_____

_____

_____

_____

Common Mistakes You Tend to Make but Want to Avoid in the Future:

_____

_____

_____

_____

_____

Types of Problems Needed for Targeted Practice:

_____

_____

_____

_____

_____

_____

Chapter 16
Understanding How to Analyze Structures of Products (Part 3): Nuclear Magnetic Resonance (NMR)

Key Concepts

NMR measures the extent to which a nucleus experiences a magnetic field. The electronic environment of a magnetic nucleus is determined by:

Proximity to a Pi Bond

Movement of electrons within a pi bond creates a magnetic field that adds to the force of an externally applied field.

How well the atom's electron cloud shields the nucleus

If there is an electron-withdrawing atom nearby, the magnetic nucleus is less shielded from the magnetic field.

Adjacent magnetic nuclei add to or subtract from the force of the externally applied field. Therefore, they affect the amount of energy released when a nucleus goes back into alignment. (In other words, they split the signal.) If there is only one type of chemically different hydrogen neighbor, the multiplicity is N + 1. If there are two types of chemically different hydrogen neighbors, they split each other's splits. In other words, the multiplicity is (N + 1) for the first type of neighbor times (N + 1) for the second type of neighbor.

Because NMR measures the environment each magnetic nucleus experiences, it is used to identify nearby functional groups, and to determine the overall structure of the molecule.

To interpret ^{1}H NMR spectra:

1. Compare the actual number of hydrogens to the number expected based on the 2N + 2 rule. If there are fewer hydrogens than expected, it means the molecule has some combination of double bonds and rings. If there are at least six fewer hydrogens than expected, it probably means the molecule has a benzene ring.

2. Look for signals in the 7–8 ppm range to determine whether the molecule has a substituted benzene ring and, if so, what characteristics that ring has. If the signal is a Multiplet Mess, it means the benzene ring is Monosubstituted. If it is a Doublet of Doublets, then the benzene ring is Disubstituted with Different groups para to each other. If it's a Singlet at approximately Seven, the ring is Substituted with the Same group in a para configuration.

3. Examine the number of signals. If there are fewer signals than expected, the structure potentially has one or more features such as a carbonyl, a *tert*-butyl group, an iso-branch, or general symmetry.

4. Look at the locations of signals to determine nearby functional groups.

5. Look at the multiplicity of each signal to determine the characteristics of immediate hydrogen neighbors.

6. Use integration data to determine the ratios of hydrogens that contributed to each signal.

7. Use coupling constants to determine which hydrogens are neighbors.

What You Need to Learn, Understand, and Apply

1. The types of information NMR provides. (page 459)

2. The general theory of NMR. (pages 459–461)

3. How to identify which atoms are chemically equivalent. Also, how to label hydrogens and carbons using signal designations such as a, b, and c. (pages 461–465)

4. Approximate values of NMR chemical shifts for general classes of ^{1}H nuclei. (page 466)

5. What causes an NMR signal to split and how to use splitting patterns to help determine the structure of a compound. (pages 466–473)

6. How to interpret and use NMR integration curves to help determine the structure of a compound. (pages 473–474)

7. How the ^{1}H NMR signal of an alcohol, amine, or carboxylic acid is affected by acid or base. (pages 474–475)

8. What NMR coupling constants are and what information they provide. (pages 475–476)

9. The advantages of using ^{13}C NMR either alone or in addition to ^{1}H NMR. Also, some of the tools that can be used to interpret ^{13}C NMR. (pages 476–478)

10. The principle of MRI. (page 478)

11. The skills needed to apply the material and avoid common errors. (pages 478–481)

Chapter Preview

The general purpose of this chapter:

_____

_____

_____

_____

_____

(Keep this purpose in mind as you read the chapter to help you tie all the concepts together into one complete picture.)

Questions to Assess Understanding	Lecture/Reading
	After filling in your lecture note outline, fold the paper over so that only the assessment question is visible. Once you can consistently answer a given question correctly on your own, place an X by that question.
_____ What information does NMR provide?	**Learning Objective 1: Know what types of information NMR provides. (page 459)** What NMR stands for: _____ Its strengths in determining the identity of a molecule: 1. _____ _____ 2. _____ _____ _____
_____ What is the general theory of NMR?	**Learning Objective 2: Know the general theory of NMR. (pages 459–461)** _____ _____ _____ _____ _____ _____

The molecular features that determine placement of a signal on an NMR chart:

_____ What determines the placement of a signal on an NMR chart?

1. _____

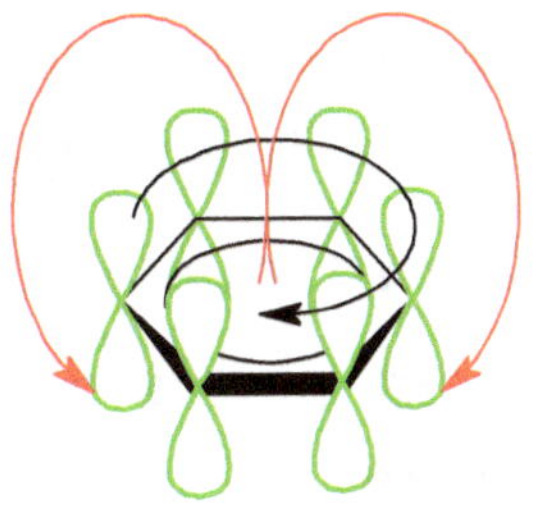

_____

_____

2. _____

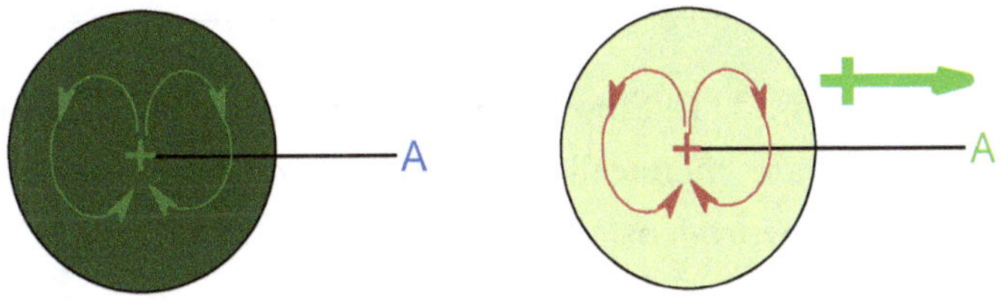

_____

_____

_____

_____

3. _____

_____

_____

_____

What the designation ppm means on an NMR chart: _____

_____

_____

Learning Objective 3:
Be able to identify chemically equivalent atoms. Also, be able to label hydrogens and carbons in a molecule using signal designations such as a, b, and c. (pages 461–465)

The number of NMR signals for a given molecule is based on the number of chemically different hydrogens or carbons the molecule has.

_____ What does it mean to say that hydrogens or carbons are chemically equivalent?

What it means to say that hydrogens or carbons are chemically equivalent:

_____

_____

_____

_____

NOTE: Single bonds allow rotation, so hydrogens attached to a given carbon are chemically equivalent. However, since a pi bond does not allow rotation, hydrogens directly attached to vinylic carbons may experience different environments.

For example,

Are in different environments, because they are affected differently by the fluorine.

$$H_2C=CHF$$

_____ What are the factors used to determine the labels a, b, c, etc. for chemically equivalent hydrogens or carbons as well as for the signals on an NMR chart?

How to determine the labels a, b, c, etc. for chemically equivalent hydrogens or carbons, as well as for the signals on an NMR chart:

1. _____

_____

_____

_____

2. _____

_____

_____

_____

Example:

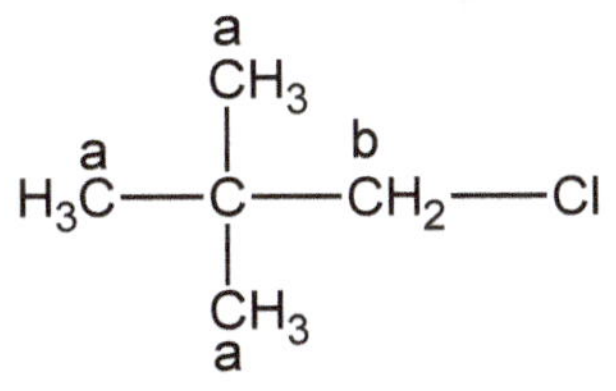

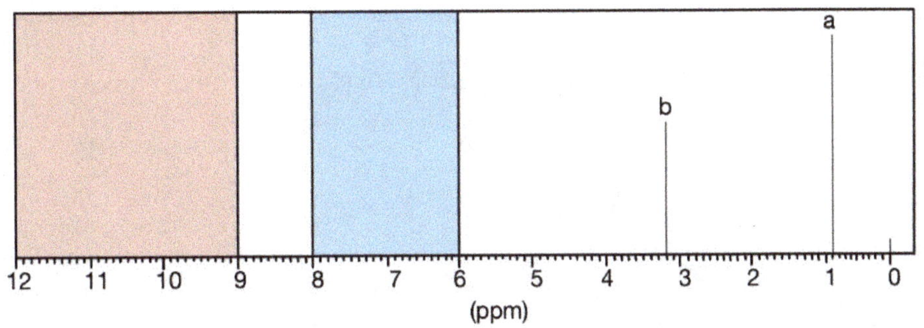

Skills Check 1

a. How many signals would you expect to see in the ^1H NMR spectrum of each of the following compounds? Assign the appropriate letter to each signal.

1. $H_3C-CH_2-CH_2-Br$

2. $H_3C-\overset{\overset{O}{\|}}{C}-CH_2-CH_2-\overset{\overset{O}{\|}}{C}-CH_3$

3. $H_3C-CH_2-O-CH_2-CH_3$

4. $H_3C-CH_2-\underset{\underset{Br}{|}}{CH}-CH_2-CH_3$

5. $H_3C-\underset{\underset{CH_3}{|}}{\overset{\overset{CH_3}{|}}{C}}-O-CH_3$

6. $H_3C-CH_2-CH_2-\overset{\overset{O}{\|}}{C}-CH_3$

7.

8. $H_3C-CH_2-\overset{\overset{O}{\|}}{C}-O-CH_3$

9.

10. $H_3C-\underset{\underset{CH_3}{|}}{CH}-CH_2-\underset{\underset{CH_3}{|}}{CH}-CH_3$

11. $H_3C-\underset{\underset{CH_3}{|}}{CH}-\overset{\overset{O}{\|}}{C}-H$

12.

13.

14.

b. How many signals would you expect to see in the ^{13}C NMR spectrum of each of the following compounds? Assign the appropriate letter to each signal.

1. $H_3C-CH_2-CH_2-Br$

2. $H_3C-\overset{\overset{\displaystyle O}{\|}}{C}-CH_2-CH_2-\overset{\overset{\displaystyle O}{\|}}{C}-CH_3$

3. $H_3C-CH_2-O-CH_2-CH_3$

4. $H_3C-CH_2-\underset{\underset{\displaystyle Br}{|}}{CH}-CH_2-CH_3$

5. $H_3C-\overset{\overset{\displaystyle CH_3}{|}}{\underset{\underset{\displaystyle CH_3}{|}}{C}}-O-CH_3$

6. $H_3C-CH_2-CH_2-\overset{\overset{\displaystyle O}{\|}}{C}-CH_3$

7. H, C=C, CH=O, H, H (alkene)

8. $H_3C-CH_2-\overset{\overset{\displaystyle O}{\|}}{C}-O-CH_3$

9. (bromobenzene with Br)

10. $H_3C-\overset{\overset{\displaystyle CH_3}{|}}{CH}-CH_2-\overset{\overset{\displaystyle CH_3}{|}}{CH}-CH_3$

11. $H_3C-\underset{\underset{\displaystyle CH_3}{|}}{CH}-\overset{\overset{\displaystyle O}{\|}}{C}-H$

12. (para-bromotoluene with Br and CH_3)

13. H_3C, H_3C, C=C, H, H (alkene)

14. H, H, C=C, H, Cl (alkene)

Learning Objective 4: Know the approximate values of NMR chemical shifts for general classes of ¹H nuclei. (page 466)

_____ What does an NMR peak at 0.9 ppm usually indicate?

_____ What does an NMR peak somewhere in the range of 7–8 ppm usually indicate?

_____ What does an NMR peak somewhere in the range of 10–12 ppm usually indicate?

Harder to flip (Downfield)   Easier to flip (Upfield)

12–10	10–8	8–6	6–4	4–2.5	2.5–1.5	1.5–0.9

| Electronegativity And Pi Bonds!!! | | More Pi Bonds!!! | Pi Bond! | Electro-negativity | Pi Bond Nearby | Minding Its Own Business |

Of course, if there are multiple features in a given molecule, they could work together to pull the signal even farther downfield than expected. Because of that, you need to be flexible when interpreting NMR spectra. However, there are three values that you can work with fairly confidently either because they are already pulled so far downfield that they can't go farther, or they are so far upfield they must not have been affected at all: 0.9 ppm (-CH₃), 7–8 ppm (benzene ring) and 10–12 ppm (aldehyde/carboxylic acid)

Learning Objective 5: Know what causes an NMR signal to split and how to use splitting patterns to help determine the structure of a compound. (pages 466–473)

The number of peaks in any particular signal is also known as the multiplicity of the signal.

The theory for why particular multiplicities are observed, and why they are seen in given ratios:

_____ What is the N + 1 rule as it relates to multiplicity?	When there is 1 chemically non-equivalent neighbor: The signal is split into a: _____ The ratios are: _____
	When there are 2 chemically non-equivalent neighbors:
_____ How do you calculate the multiplicity of an NMR signal if the hydrogen has neighbors on each side that are not chemically equivalent to that hydrogen, but are chemically equivalent to each other?	The signal is split into a: _____ The ratios are: _____
	When there are 3 chemically non-equivalent neighbors:
_____ How do you calculate the multiplicity of an NMR signal if the hydrogen has neighbors on each side that are not chemically equivalent to that hydrogen and are also not chemically equivalent to each other?	The signal is split into a: _____ The ratios are: _____ The N + 1 rule: _____ _____ _____

The multiplicity of the signal is abbreviated as s for singlet, d for doublet, t for triplet, q for quartet, and m for multiplet.

What happens if a hydrogen has neighbors on both sides? It depends on whether the neighbors are chemically equivalent to each other.

If the neighbors are chemically equivalent to each other: _____

_____

_____

_____

If the neighbors are not chemically equivalent to each other: _____

_____

_____

_____

Examples:

$BrCH_2CH_2CH_2Br$

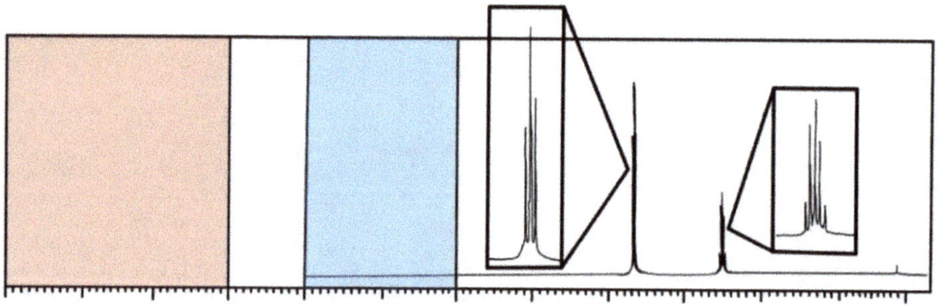

$CH_3CH_2CH_2I$

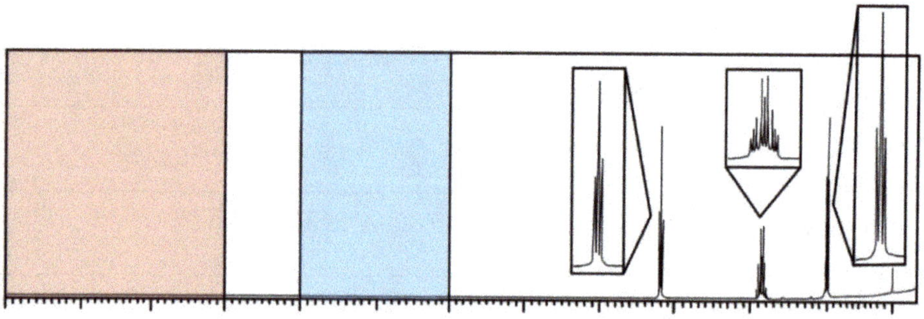

Skills Check 2

a. Write the number of ^{1}H NMR signals for each example. Then, assign the appropriate letter and multiplicity to each signal.

$$H_3C-\underset{\underset{F}{|}}{CH}-CH_3 \qquad\qquad H_3C-CH_2-CH_2-F$$

$$F-CH_2-CH_2-F$$

b. For each molecule shown in **Skills Check 1a**, write the multiplicity of each signal.

(Be aware that the splitting of two non-identical hydrogens bonded to the same sp^2 hybridized carbon (called germinal coupling) is often too minimal to be easily seen.)

Characteristic patterns for signals of aryl hydrogens based on selected structures of aromatic compounds.

Monosubstituted

_____ What is the characteristic NMR pattern for aryl hydrogens of a benzene ring when the ring is mono-substituted?

Multiplicity and signals based on structure:

Monosubstituted = **M**ess (or **M**ultiplet)

Example: Toluene

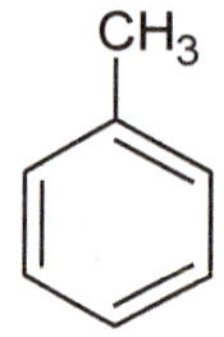

CH₃

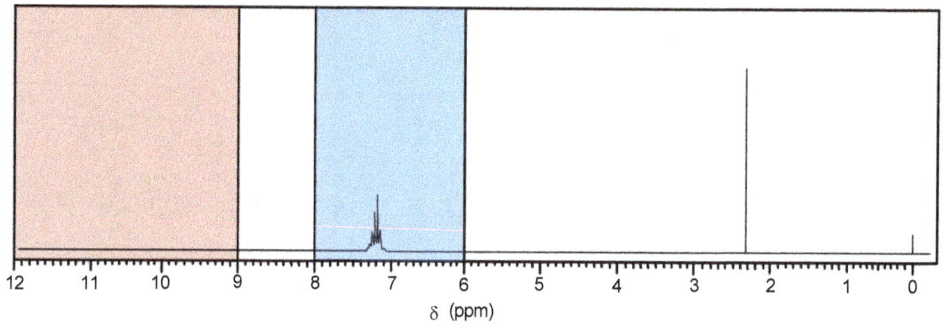

δ (ppm)

Disubstituted Different (Para)

Multiplicity and signals based on structure

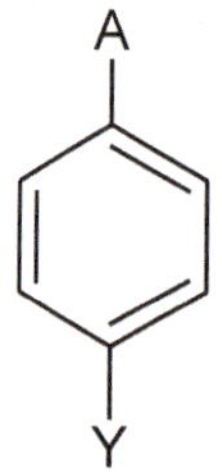

_____ What is the characteristic NMR pattern for aryl hydrogens of a benzene ring that is disubstituted with different groups that are para to each other?

A

Y

Disubstituted **D**ifferent = **D**oublet of **D**oublets

Example: para-chlorotoluene

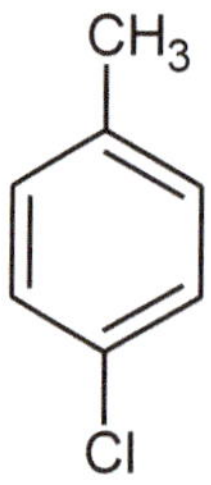

CH₃

Cl

Substituted Same (Para)

Multiplicity and signals based on structure

<u>S</u>ubstituted <u>S</u>ame = <u>S</u>inglet at <u>S</u>even

Example: para-xylene

_____ What is the characteristic NMR pattern for aryl hydrogens of a benzene ring that is substituted with the same groups and are para to each other?

Learning Objective 6: Know how to interpret and use integration curves to help determine the structure of a compound. (pages 473–474)

An integration line on an NMR chart indicates relative area under each signal. The more hydrogens that contribute to a signal, the greater the area. Use integration lines to calculate the ratios of hydrogens contributing to each signal.

_____ What do integration lines on an NMR chart tell you about the structure of a molecule?

Example:

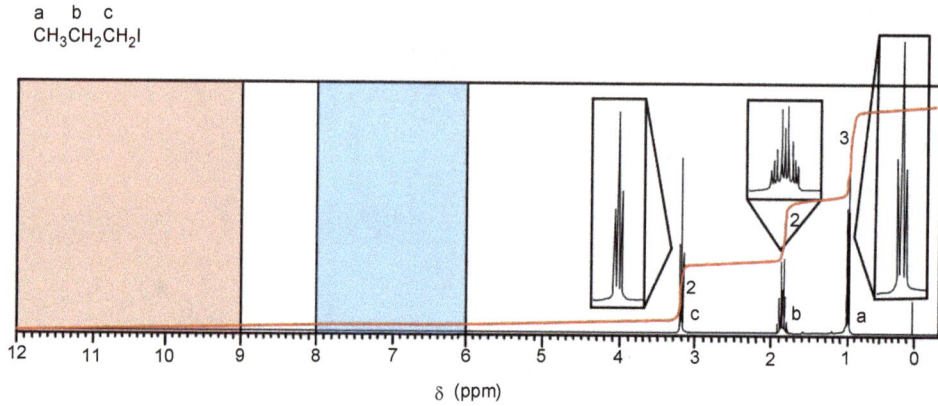

Skills Check 3

Other than signal placement, how could you distinguish among the ^1H NMR spectra of the following compounds?

Learning Objective 7: Know how the ¹H signal of an alcohol, amine, or carboxylic acid is affected by acid or base. (pages 474–475)

The issue: _____

_____

What it means for the spectrum if acid or base is present:

_____

_____

_____

_____

What it means for the spectrum if D₂O is also present:

_____

_____

_____

_____

_____ How is the signal of a hydrogen attached to nitrogen or oxygen affected if the sample is contaminated with trace amounts of acid or base?

This can be used to easily identify alcohols or amines.

Example: Ethanol

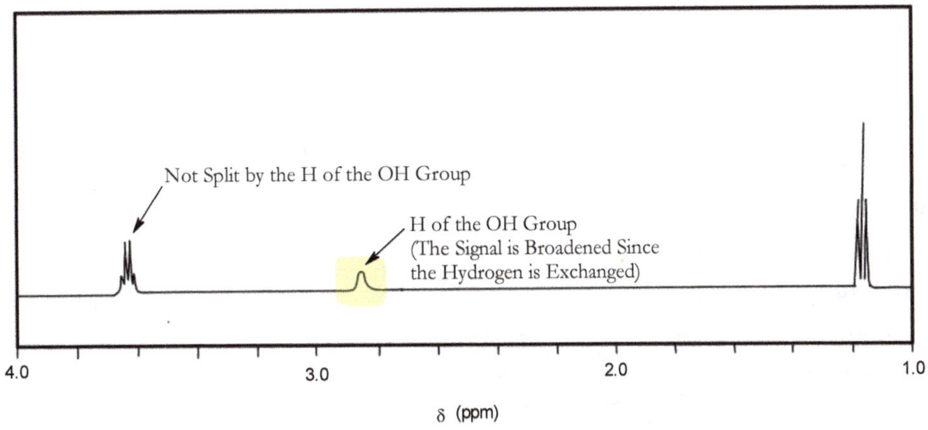

Not Split by the H of the OH Group

H of the OH Group
(The Signal is Broadened Since the Hydrogen is Exchanged)

δ (ppm)

4.0   3.0   2.0   1.0

Learning Objective 8: Know what NMR coupling constants are and what information they provide. (pages 475–476)

The definition of coupling constant: _____

_____

_____

What J_{ab} and J_{ba} refer to in the notation for coupling constants:

_____

_____

The information that can be gained from examining coupling constants:

_____

_____

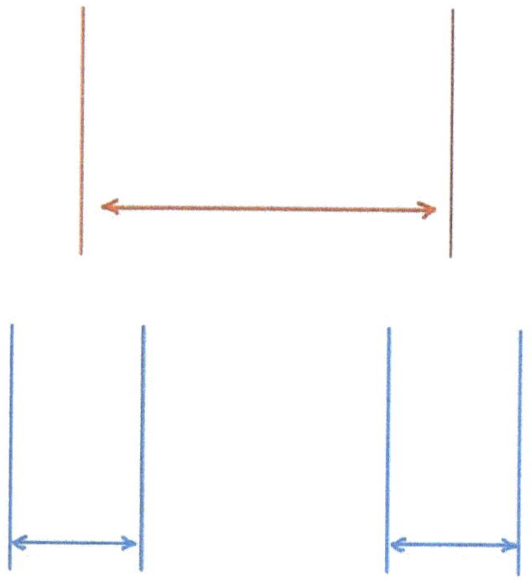

_____ What is the definition of a coupling constant?

_____ What information can be gained from examining coupling constants?

How coupling constants can affect the signal splitting that is observed when two non-chemically equivalent neighbors are present.

How it can be used to determine the difference between a quartet and a doublet of doublets:

_____

_____

_____

Learning Objective 9: Know the advantages of using ^{13}C NMR either alone or in addition to ^{1}H NMR. Also, know some of the tools that can be used to interpret ^{13}C NMR. (pages 476–478)

Some Advantages of ^{13}C NMR:

1. _____

_____

_____

2. _____

_____

_____

General Information about ^{13}C NMR:

APPROXIMATE Ranges:

150 +	150–100	100–50	50–0
Electronegativity And Pi Bonds!!!	Pi Bonds !	Electro-negativity	Minding Its Own Business

_____ What are some advantages of ^{13}C NMR over ^{1}H NMR?

_____ In general, what types of carbons are likely to give a signal between 0 and 50 ppm in ^{13}C NMR?

_____ between 50 and 100 ppm in ^{13}C NMR?

_____ between 100 and 150 ppm in ^{13}C NMR?

_____ greater than 150 ppm in ^{13}C NMR?

Integration is not used for ^{13}C NMR. Also, signals in ^{13}C NMR are not usually split because the probability that the neighboring carbon will be a ^{13}C isotope is incredibly small. However, if an NMR is run in spin-coupled mode, the attached hydrogens can be used to split the signal. A coupled NMR indicates how many hydrogens are attached to a given carbon (rather than to the carbon neighbor).

Example: 2-butanol

Not Spin Coupled:

Spin Coupled:

Learning Objective 10: Know the principle of MRI. (page 478)

How MRIs create images: _____

_____

_____

Learning Objective 11: Gain the skills needed to apply the material and avoid common errors. (pages 478–481)

There are seven tools you can use to interpret an NMR chart. For any given problem, you may use one, more than one, or all seven. They are:

_____ What is the general principle for how MRIs create images?

1. _____

_____

_____

2. _____

_____

_____

3. _____

_____

_____

4. _____

_____

_____

5. _____

_____

6. _____

_____

_____

7. _____

_____

_____

_____ What are the tools you can use to interpret an NMR chart?

The first five factors alone will usually help you figure out the structure of the molecule. The fourth factor (signal position) can also be really useful, but you have to remember that if multiple factors are present, the signal doesn't stay in the range you might think, so you have to be flexible. The most unambiguous signals in ^{1}H NMR are those in the ranges of 10–12, 7–8, and 0.9–1.0 ppm.

Additional pieces of information to help you determine the structure of a molecular based on an NMR spectrum:

Multiplet "Sammiches" (Sandwiches): _____

_____

_____

Triplet/quartet combinations: _____

Doublets: _____

Interpreting Multiplicities around 7 ppm:

Multiplet: _____

Doublet of Doublets:_____

Singlet: _____

Explanations of How to Work the
Learn to Analyze and Apply Problems:

The methodical process for determining the identity of a molecule based on a combination of its molecular formula, MS, IR and/or NMR data:

1. _____

_____

2. _____

_____

3. _____

_____

4. _____

_____

5. _____

_____

When selecting multiple choice answers for questions related to instrumentation data:

_____

_____

_____

_____

_____

Learn to Analyze and Apply

Learn to Analyze and Apply 1: Determine the structure of a molecule based on its NMR data

Determine the molecular structure of the compound that produced the following NMR data:

Molecular Formula: $C_6H_{12}O_2$

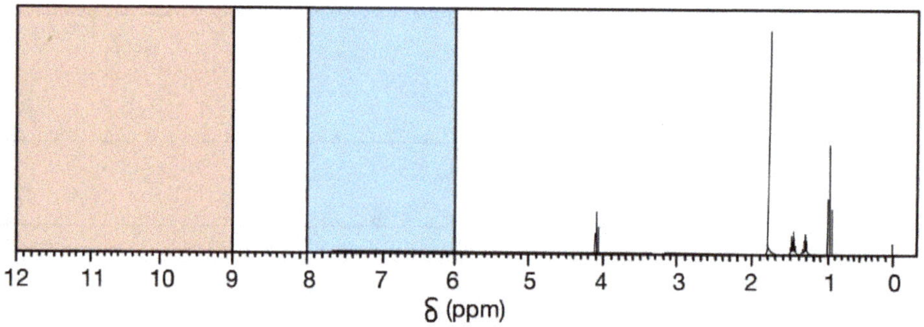

STEP 1:
Determine the number of pi bonds/rings
based on the 2N + 2 rule.

# of pi bonds/rings: _____

STEP 2:
If you find that hydrogens are "missing" based on the
2N + 2 rule, determine the type of pi bond or ring
based on characteristic placement of signals in the ranges
of 6–8 and 9–12 ppm. Confirm your hypothesis
by examining other regions.

Interpretation of any signals in the 6–8 ppm region: _____

Interpretation of any signals in the 9–12 ppm region: _____

_____

STEP 3:
Keep a running tally of the number of carbons and hydrogens
that remain after subtracting known structures.

(Subtract the number of carbons and hydrogens in a substituted benzene ring, the carbon, hydrogen and oxygen for an aldehyde, the carbon, hydrogen and two oxygens of a carboxylic acid or ester, etc.)

_____

STEP 4:
<mark>Compare the number of remaining NMR signals to the number of carbons.</mark>

Number of carbons without hydrogen, potential number of iso branches, and/or degree of symmetry:

_____

_____

STEP 5:
As necessary, use <mark>multiplicity, signal location, and integration values</mark> to rule potential structures in/out.

Helpful clues related to multipilicities: _____

_____

_____

If signals are present at approximately 2.5 or 4, what they tell you about the structure:

_____

Helpful information related to integration values: _____

Putting everything together, the structure of the molecule is:

Learn to Analyze and Apply 2: Determine the Identity of a Molecule Based on its Molecular Formula, IR, and NMR data

Determine the molecular structure of a compound with the molecular formula of $C_{11}H_{14}O_2$ that produced the following IR and NMR data:

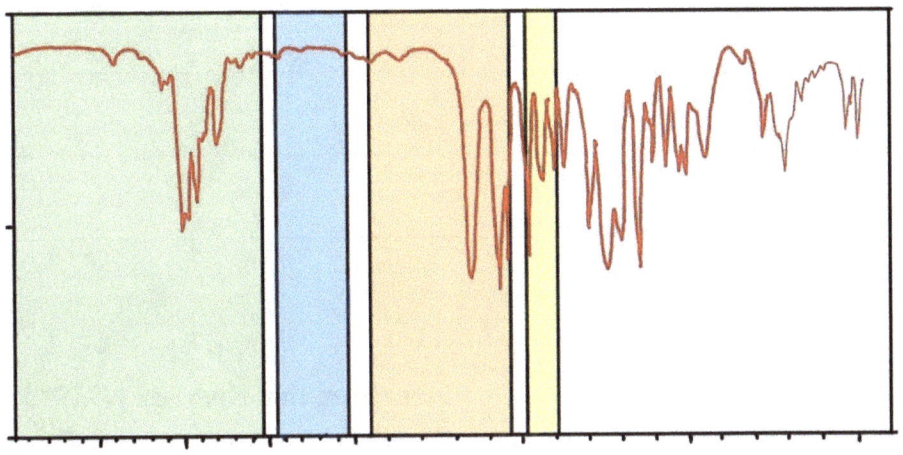

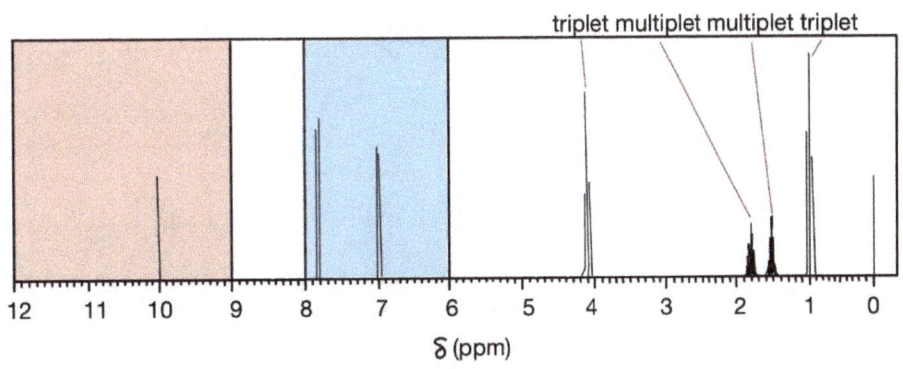

triplet multiplet multiplet triplet

δ (ppm)

STEP 1:
Determine the number of pi bonds/rings
based on the 2N + 2 rule.

_____

# of pi bonds/rings: _____

_____

STEP 2a:
Interpret the clearest IR data
Look for easily identifiable bands in the 1700–3500 cm^{-1} range.
When possible, verify your interpretation by looking for supporting bands.

Interpretation of bands in the 1700–3500 cm^{-1} region: _____

_____

_____

_____

_____

STEP 2b:
Interpret the clearest NMR data
Interpret signals in the ranges of 6–8 and 9–12 ppm.

Interpretation of signals in the 6–8 ppm region: _____

_____

Interpretation of signals in the 9–12 ppm region: _____

_____

STEP 2c
Verify agreement between IR and NMR data.

_____

_____

_____

_____

STEP 3:

Keep a <mark>running tally</mark> of the number of carbons and hydrogens that remain after subtracting known structures.

_____

STEP 4:

<mark>Compare the number of remaining NMR signals to the number of carbons.</mark>

Number of carbons without hydrogen, potential number of iso- branches, and/or degree of symmetry:

_____

_____

STEP 5:

As necessary, use <mark>signal location, multiplicity, and integration values</mark> to rule potential structures in/out.

Helpful clues related to multipilicities: _____

_____

If signals are present at approximately 2.5 or 4, what they tell you about the structure:

_____

_____

Helpful information related to integration values: _____

Putting everything together, the structure of the molecule is:

Learn to Analyze and Apply 3: Determine the Identity of a Molecule Based on its MS, IR, and NMR data

Determine the molecular structure of a compound that produced the following MS, IR, and NMR data:

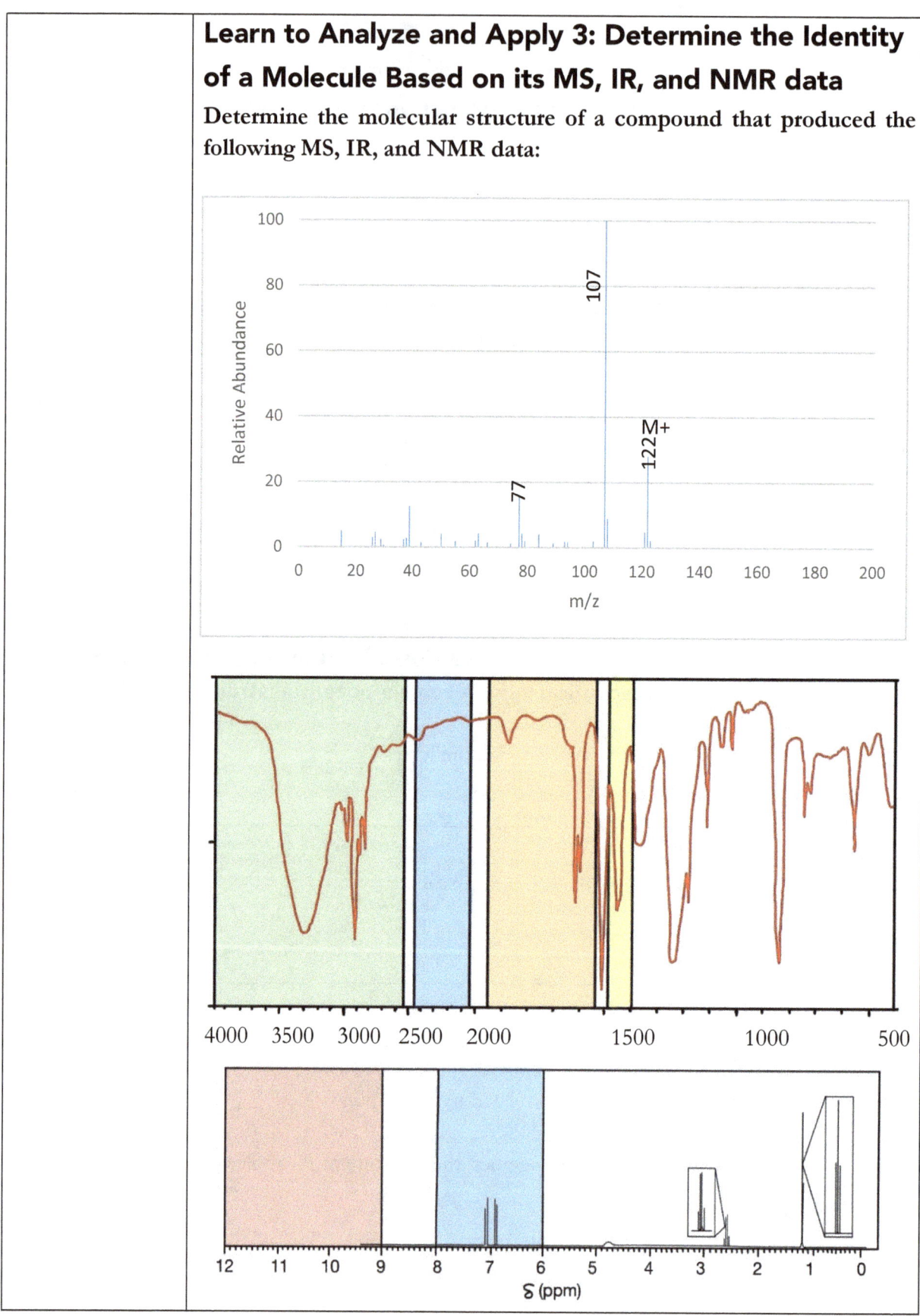

STEP 1:
Interpret the clearest MS data
(Look for evidence of Cl, Br, or N)

Is there evidence of Cl, Br, or N? If so, which is present? _____

STEP 2a:
Interpret the clearest IR data.

Look for easily identifiable bands in the 1700–3500 cm^{-1} range. When possible, verify your interpretation by looking for supporting bands.

Interpretation of bands in the 1700–3500 cm^{-1} region: _____

_____

_____

_____

_____

STEP 2b:
Interpret the clearest NMR data
Interpret signals in the ranges of 6–8 and 9–12 ppm.

Interpretation of signals in the 6–8 ppm region: _____

Interpretation of signals in the 9–12 ppm region: _____

STEP 2c:
Verify agreement between MS, IR, and NMR data.

_____

_____

_____

_____

STEP 3:

Keep a <mark>running tally</mark> of the number of carbons and hydrogens that remain after subtracting known structures.

_____

STEP 4:

<mark>Compare the number of remaining NMR signals to the number of carbons.</mark>

Number of carbons without hydrogen, potential number of iso branches, and/or degree of symmetry:

_____

_____

_____

STEP 5:

As necessary, use <mark>multiplicity, signal location, and integration values</mark> to rule potential structures in/out.

Helpful clues related to multipilicities: _____

_____

_____

If signals are present at approximately 2.5 or 4, what they tell you about the structure:

_____

_____

Helpful information related to integration values: _____

Putting everything together, the structure of the molecule is:

Integrate Skills

Predict NMR Signals

1. Draw the indicated structure, label the hydrogens as a, b, c, etc., write the predicted number of ^1H NMR signals, and then predict the approximate chemical shift range (in ppm), multiplicity (singlet, double, etc.) and integration ratios for each signal: (For example, a: 1.5–2.5 ppm (m) 3H, b: 2.5–4 ppm (d) 2H, etc.)

a. pentanoic acid
    Structure:

    Number of signals: _____

    Predicted multiplicity, integration ratio, and approximate chemical shift range for each signal (include identifying a, b, c, etc. labels for each):

    _____

    _____

    _____

b. isopentyl ether

    Structure:

    Number of signals: _____

    Predicted multiplicity, integration ratio, and approximate chemical shift range for each signal (include identifying a, b, c, etc. labels for each):

    _____

    _____

    _____

c. butyl acetate

Structure:

Number of signals: _____

    Predicted multiplicity, integration ratio, and approximate chemical shift range for each signal (include identifying a, b, c, etc. labels for each):

_____

_____

_____

d. 4-methylpentanal

Structure:

Number of signals: _____

    Predicted multiplicity, integration ratio, and approximate chemical shift range for each signal (include identifying a, b, c, etc. labels for each):

_____

_____

_____

e. propyl benzene

Structure:

Number of signals: _____

    Predicted multiplicity, integration ratio, and approximate chemical shift range for each signal (include identifying a, b, c, etc. labels for each):

_____

_____

_____

f. *p*-bromotoluene

Structure:

Number of signals: _____

    Predicted multiplicity, integration ratio, and approximate chemical shift range for each signal (include identifying a, b, c, etc. labels for each):

_____

_____

_____

g.

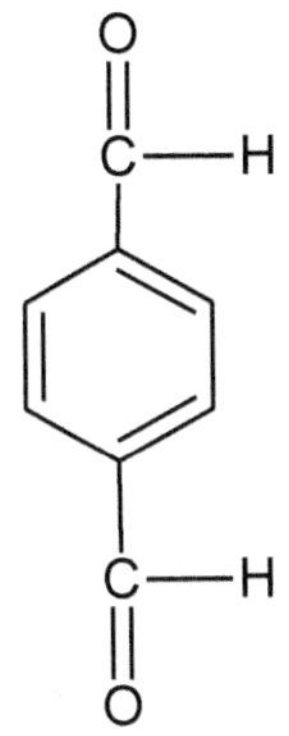

Number of signals: _____

Predicted multiplicity, integration ratio, and approximate chemical shift range for each signal (include identifying a, b, c, etc. labels for each):

_____

_____

_____

h. p-isopropylbenzaldehyde

Structure:

Number of signals: _____

Predicted multiplicity, integration ratio, and approximate chemical shift range for each signal (include identifying a, b, c, etc. labels for each):

_____

_____

_____

i. 3-methylbutanal

Structure:

Number of signals: _____

Predicted multiplicity, integration ratio, and approximate chemical shift range for each signal (include identifying a, b, c, etc. labels for each):

_____

_____

_____

Chapter 16

Match the NMR data to the Structure

2. Circle the structure that correlates with the ^{1}H NMR data:

a.

1H   3H

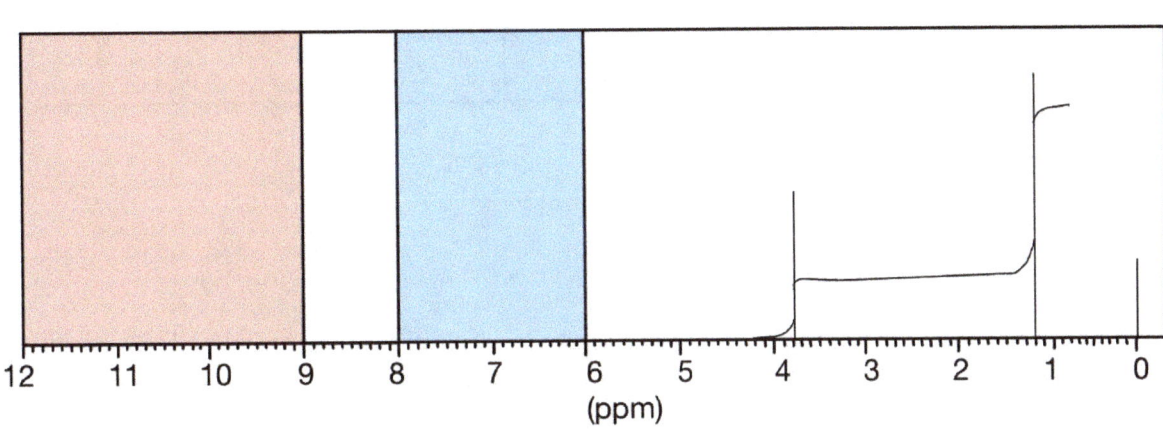

(ppm)

Choices for a:

$$H_3C-\overset{\overset{\displaystyle CH_3}{|}}{CH}-O-\overset{\overset{\displaystyle O}{||}}{C}-CH_3$$

$$H_3C-\overset{\overset{\displaystyle CH_3}{|}}{CH}-\overset{\overset{\displaystyle O}{||}}{C}-O-CH_3$$

$$H_3C-\overset{\overset{\displaystyle CH_3}{|}}{\underset{\underset{\displaystyle CH_3}{|}}{C}}-O-\overset{\overset{\displaystyle O}{||}}{C}-CH_3$$

$$H_3C-\overset{\overset{\displaystyle CH_3}{|}}{\underset{\underset{\displaystyle CH_3}{|}}{C}}-\overset{\overset{\displaystyle O}{||}}{C}-O-CH_3$$

$$H_3C-\overset{\overset{\displaystyle CH_3}{|}}{\underset{\underset{\displaystyle CH_3}{|}}{C}}-\overset{\overset{\displaystyle O}{||}}{C}-CH_3$$

$$H_3C-\overset{\overset{\displaystyle O}{||}}{C}-O-CH_3$$

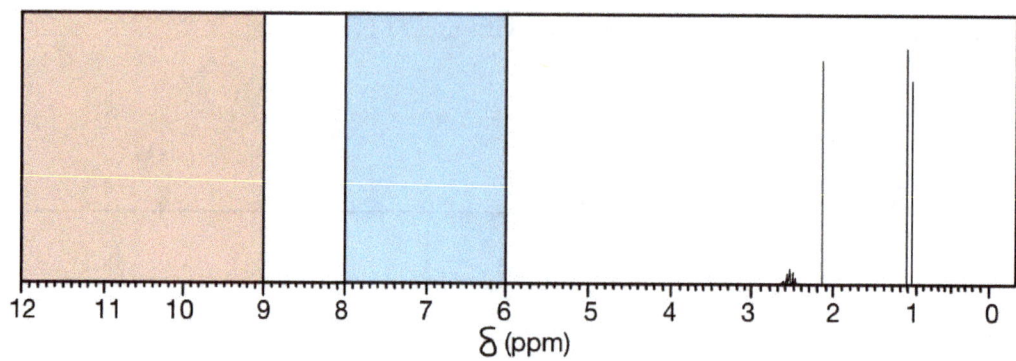

b.

Choices for b:

H₃C—CH₂—CH₂—C(=O)—CH₂—CH₂—CH₃

H₃C—CH₂—C(=O)—CH₃

H₃C—CH(CH₃)—CH₂—C(=O)—O—CH₂—CH₃

H₃C—CH(CH₃)—C(=O)—O—CH₂—CH₃

H₃C—CH₂—CH₂—C(=O)—O—CH₂—CH₃

H₃C—CH(CH₃)—C(=O)—O—CH₃

H₃C—CH₂—CH₂—O—C(=O)—CH₂—CH₃

H₃C—CH₂—C(=O)—O—CH₃

H₃C—CH₂—CH₂—C(=O)—CH₂—CH₂—CH₃

H₃C—CH₂—O—C(=O)—CH₃

H₃C—CH₂—CH₂—CH₂—CH₂—C(=O)—OH

H₃C—CH(CH₃)—C(=O)—CH₃

c.

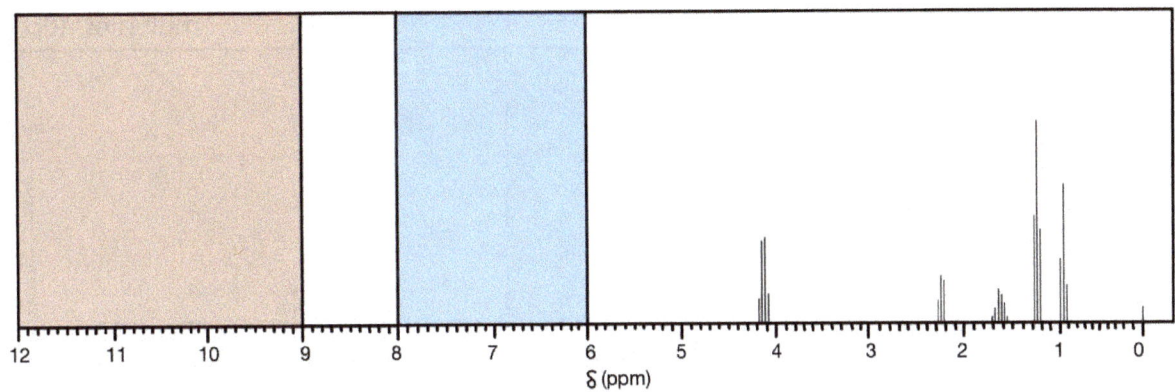

δ (ppm)

Choices for c:

$$H_3C-\underset{\underset{CH_3}{|}}{CH}-CH_2-\overset{\overset{O}{\|}}{C}-O-CH_2-CH_3$$

$$H_3C-CH_2-CH_2-\overset{\overset{O}{\|}}{C}-O-CH_2-CH_3$$

$$H_3C-CH_2-CH_2-O-\overset{\overset{O}{\|}}{C}-CH_2-CH_3$$

$$H_3C-CH_2-CH_2-CH_2-CH_2-\overset{\overset{O}{\|}}{C}-OH$$

$$H_3C-\underset{\underset{CH_3}{|}}{CH}-CH_2-O-\overset{\overset{O}{\|}}{C}-CH_2-CH_3$$

$$H_3C-\underset{\underset{CH_3}{|}}{CH}-O-\overset{\overset{O}{\|}}{C}-CH_2-CH_3$$

$$H_3C-\underset{\underset{CH_3}{|}}{CH}-\overset{\overset{O}{\|}}{C}-O-CH_2-CH_3$$

d.

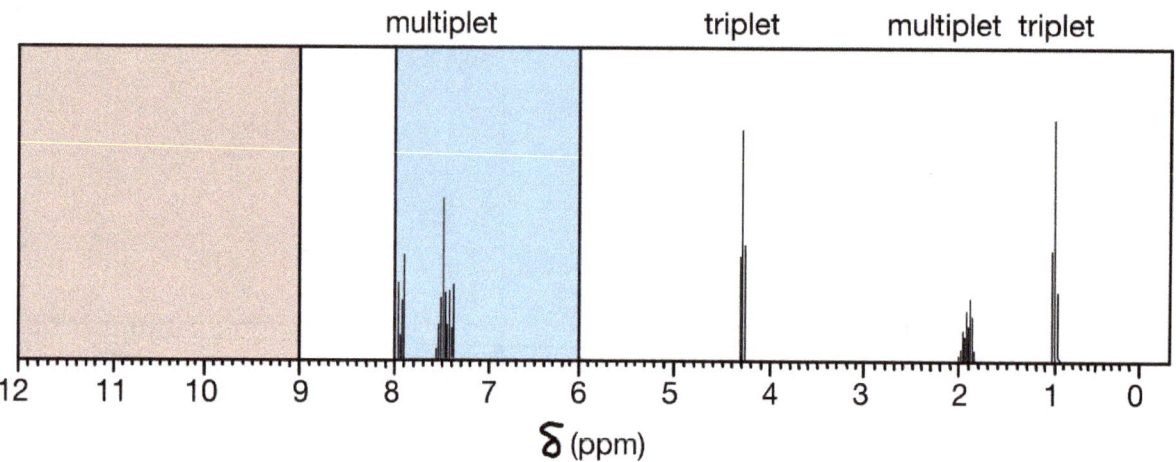

δ (ppm)

Answer choices for d:

e.

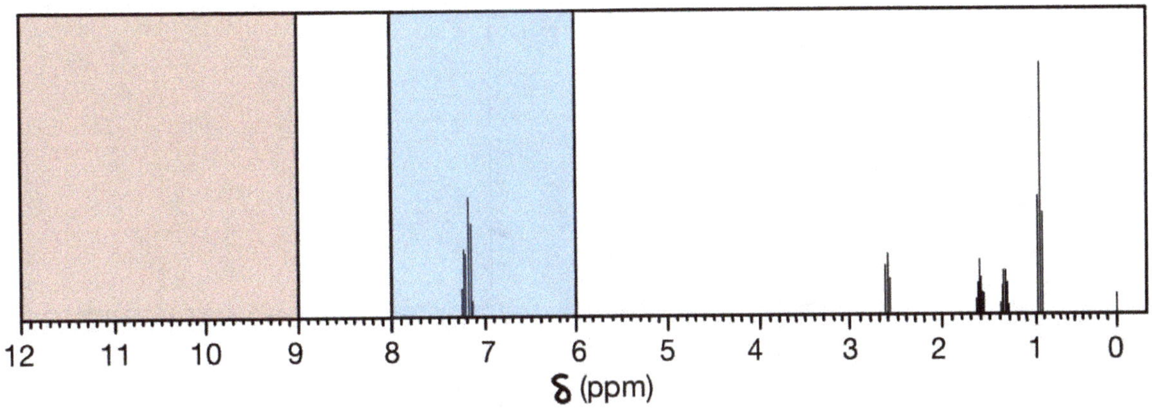
12
11
10
9
8
7
6
5
4
3
2
1
0
δ (ppm)

Answer choices for e:

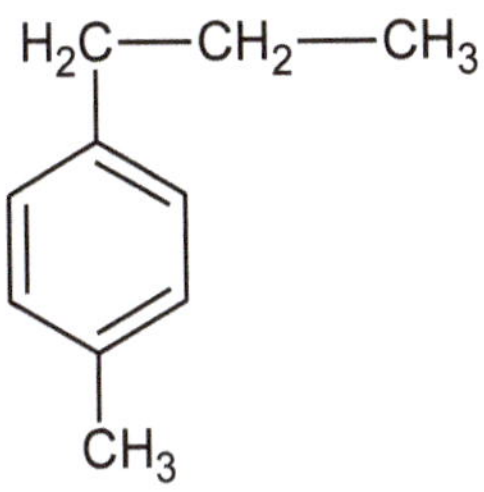
H2C—CH2—CH3
CH3

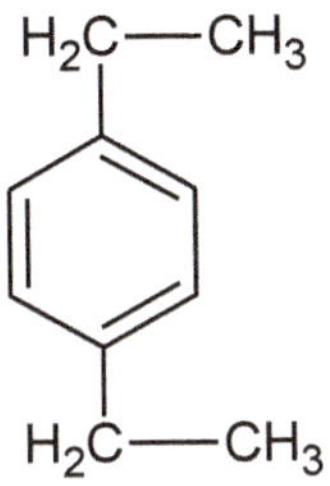
H2C—CH3
H2C—CH3

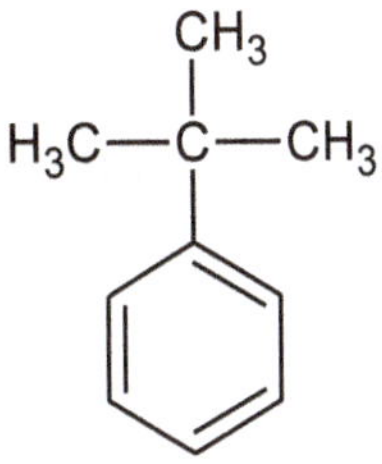
CH3
H3C—C—CH3

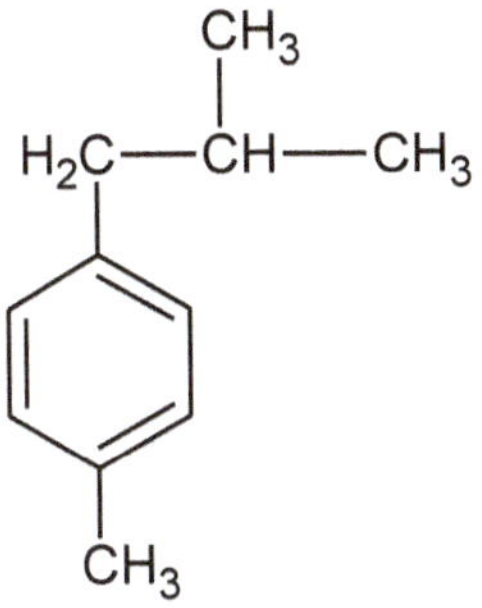
CH3
H2C—CH—CH3
CH3

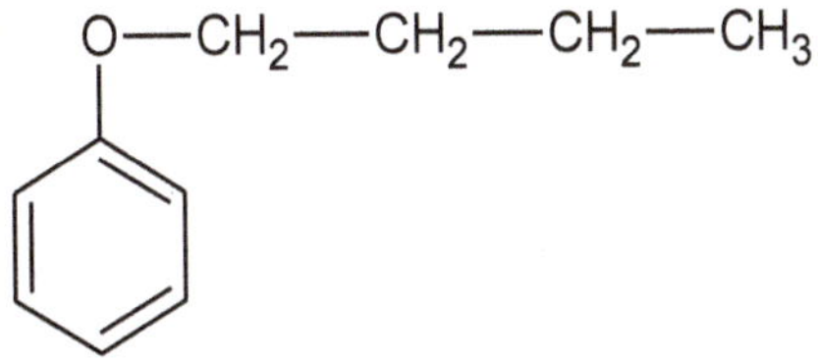
O—CH2—CH2—CH2—CH3

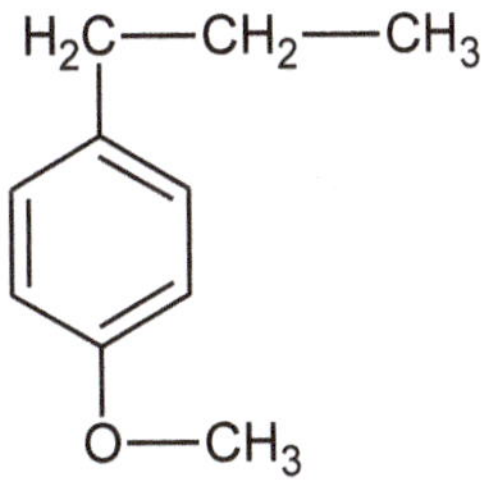
H2C—CH2—CH3
O—CH3

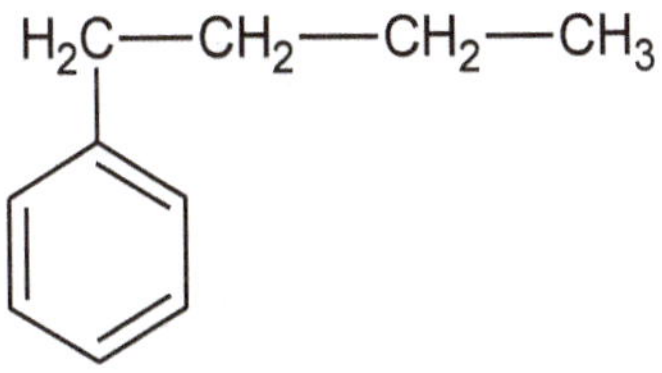
H2C—CH2—CH2—CH3

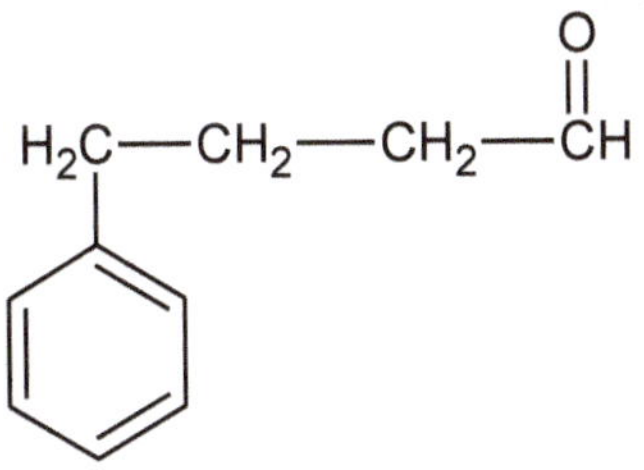
O
H2C—CH2—CH2—CH

f.

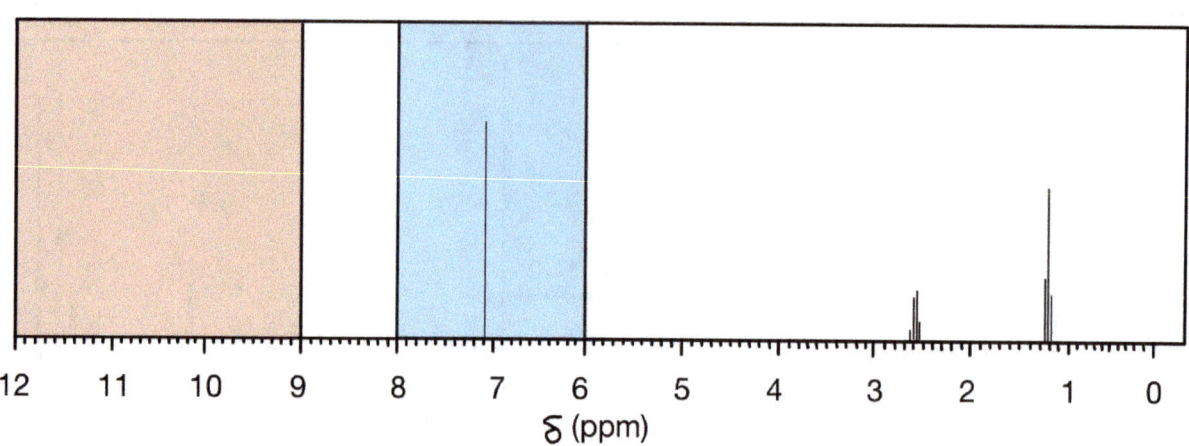

Answer choices for f:

g.

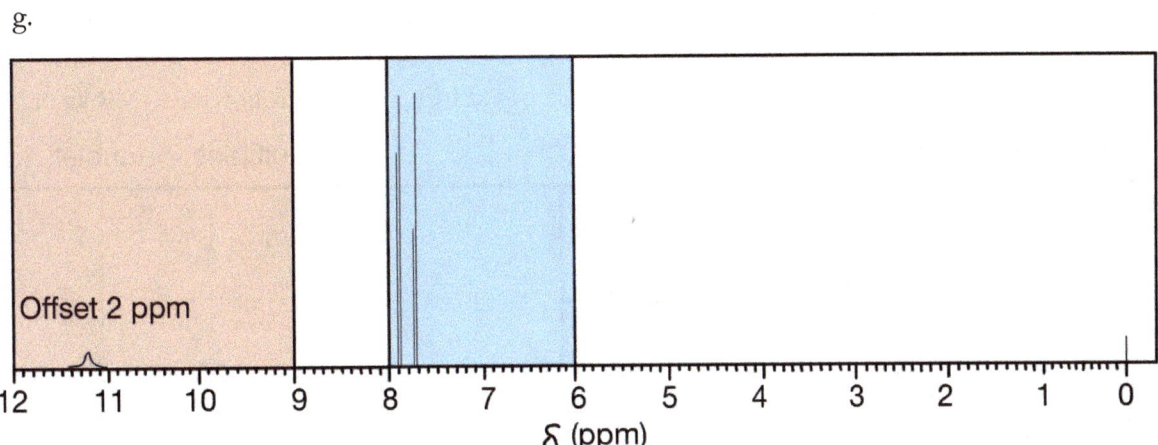

Offset 2 ppm

δ (ppm)

Answer Choices for g:

H_2C—CH_2—CH_2—$\overset{\overset{\displaystyle O}{\|}}{C}H$

O—CH_2—CH_2—CH_2—CH_3

CH_3

Br

H_2C—CH_3

Br

$\overset{\overset{\displaystyle O}{\|}}{C}$—OH

$\overset{\overset{\displaystyle O}{\|}}{C}$—OH

$\overset{\displaystyle O}{\|}$
C—OH

$\overset{\overset{\displaystyle O}{\|}}{C}$—OH

Br

Determine Molecular Structure Based on the Molecular Formula and ¹H NMR Chart

3.

    a. A compound with the molecular formula of $C_4H_8O_2$ gives the following NMR signals:

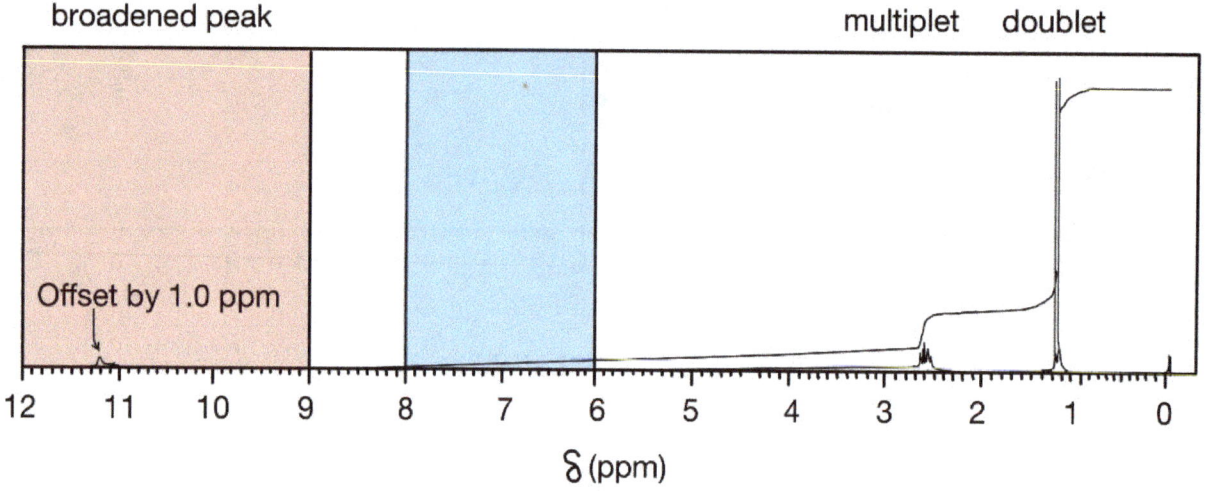

What is the structure of the compound?

b. A compound with the molecular formula of $C_{12}H_{18}$ gives the following NMR signals:

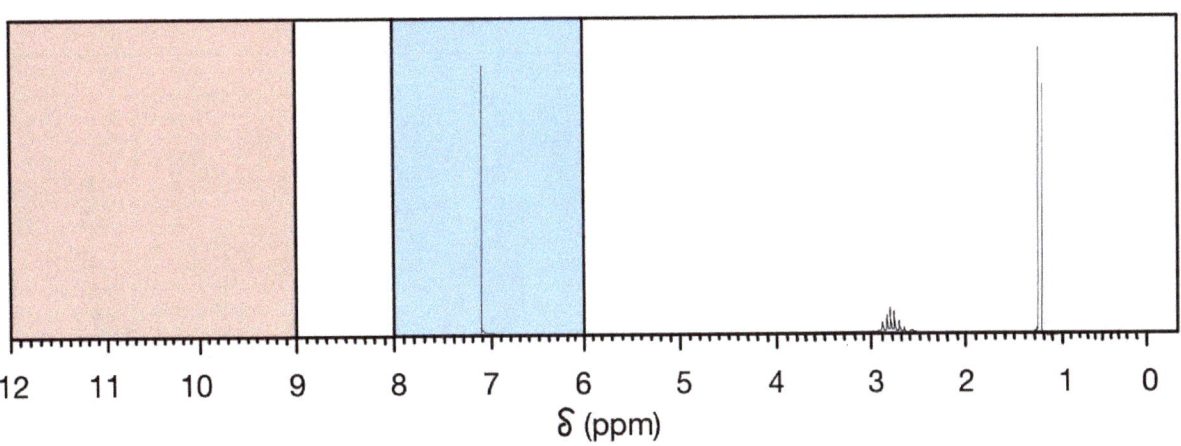

What is the structure of the compound?

c. A compound with the molecular formula of $C_8H_{10}O$ gives the following NMR signals:

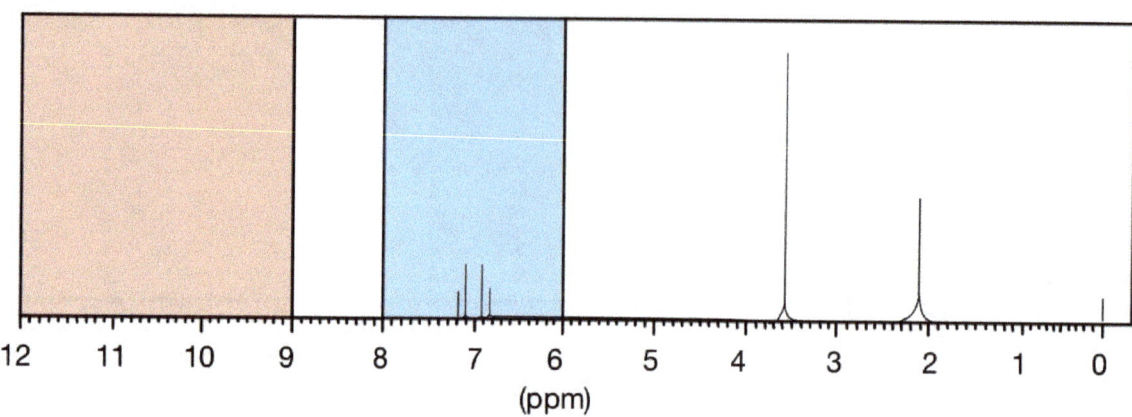

What is the structure of the compound?

Chapter 16

d. A compound with the molecular formula of $C_5H_{11}Br$ gives the following NMR signals:

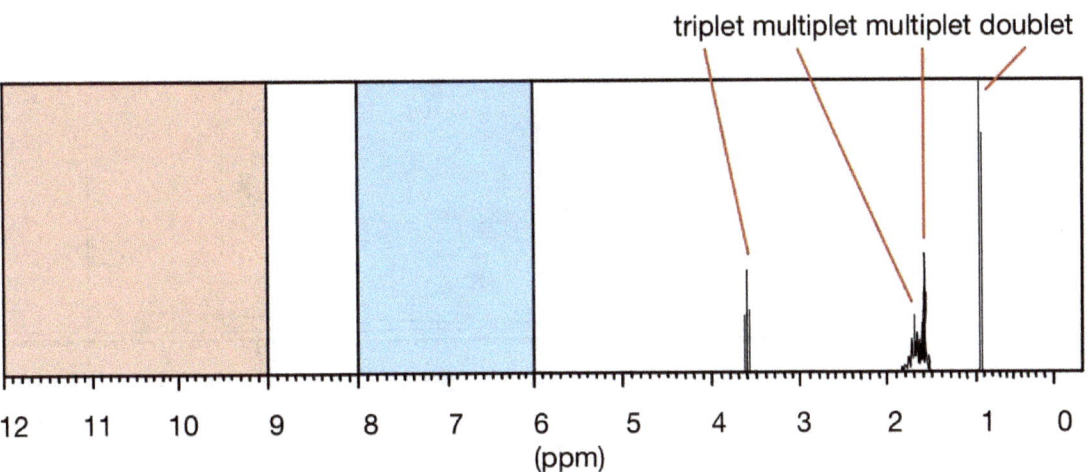

What is the structure of the compound?

e. A compound with the molecular formula of $C_9H_{18}O$ gives the following NMR signals:

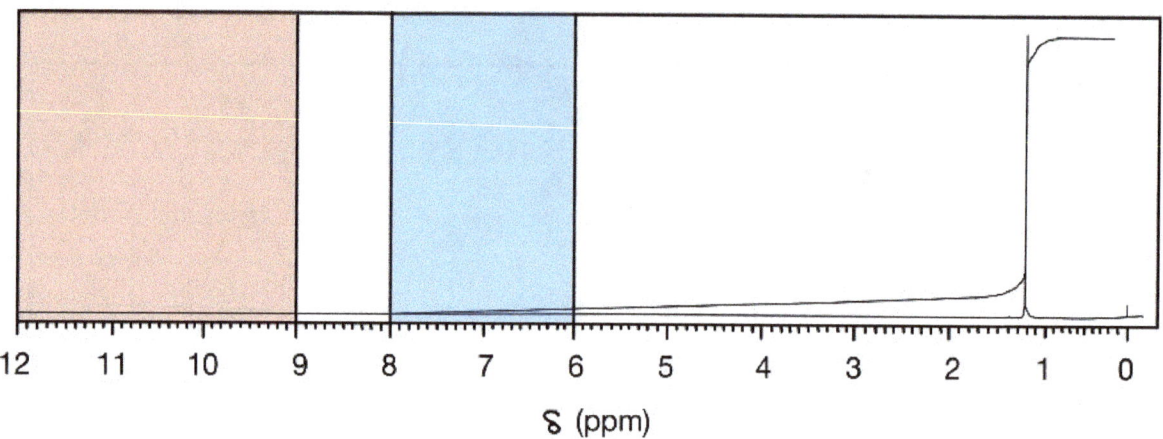

What is the structure of the compound?

f. A compound with the molecular formula of C_8H_8O gives the following NMR signals:

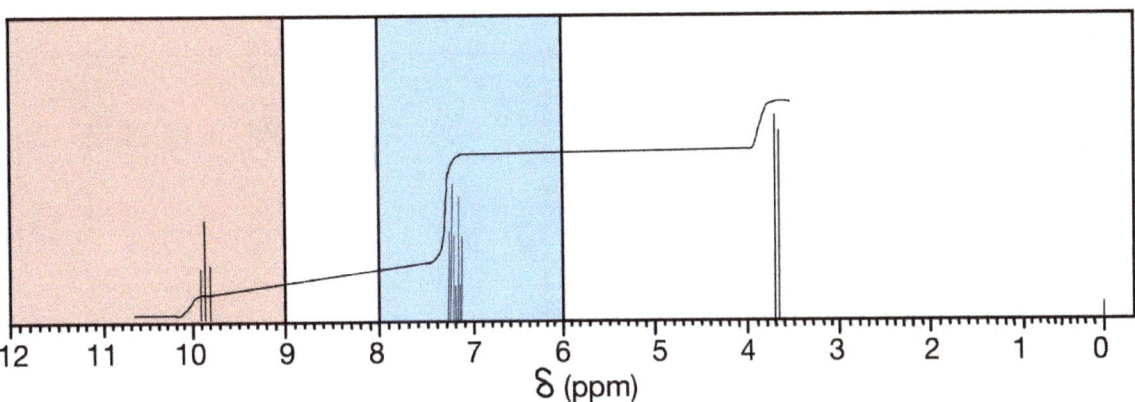

What is the structure of the compound?

g. A compound with the molecular formula of $C_6H_{10}O_2$ gives the following NMR signals:

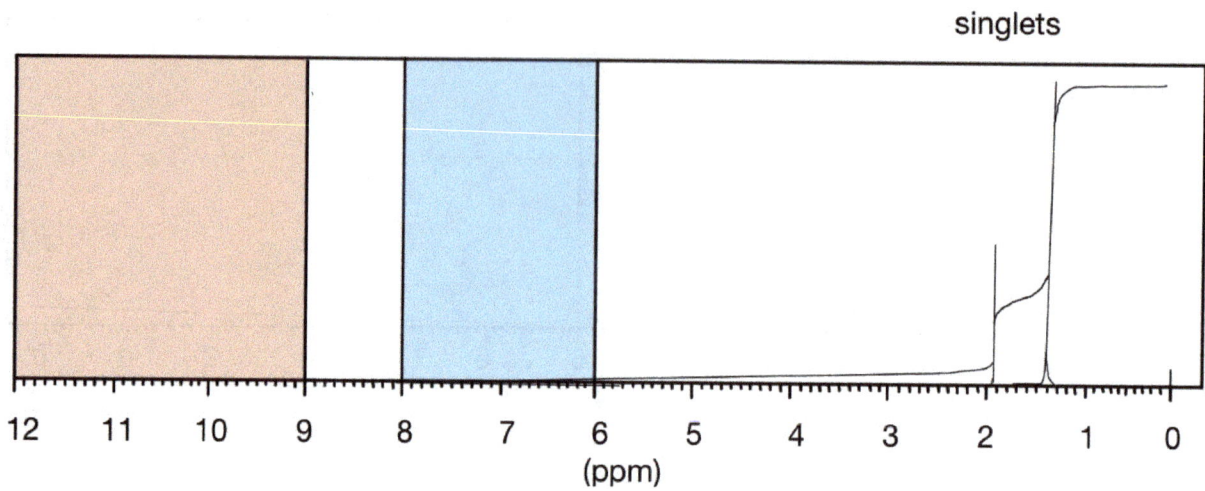

What is the structure of the compound?

Determine Molecular Structure Based on a Molecular Formula and Written Description of ¹H NMR data

4.

a. A compound with the molecular formula of C_8H_8O gives the following NMR signals:

d δ 3.7 (2H), m δ 7.2 (5H), t δ 9.7 (1 H)

What is the structure of the compound?

b. A compound with the molecular formula of $C_6H_{12}O_2$ gives the following NMR signals:

d δ 0.95 (6H), m δ 1.9 (1H), s δ 2.1 (3H), d δ 3.88 (2H)

What is the structure of the compound?

c. A compound with the molecular formula of C_5H_{10} gives the following NMR signals:
t δ 1.05 (3H), q δ 2.8 (2H)

What is the structure of the compound?

d. A compound with the molecular formula of $C_6H_{12}O_2$ gives the following NMR signals:
t δ 0.95, m δ 1.4, m δ 1.6, s δ 2.05, t δ 4.07

What is the structure of the compound?

Interpreting C-13 NMR

5. Match each structure to its ^{13}C NMR Data within the group:

Group 1

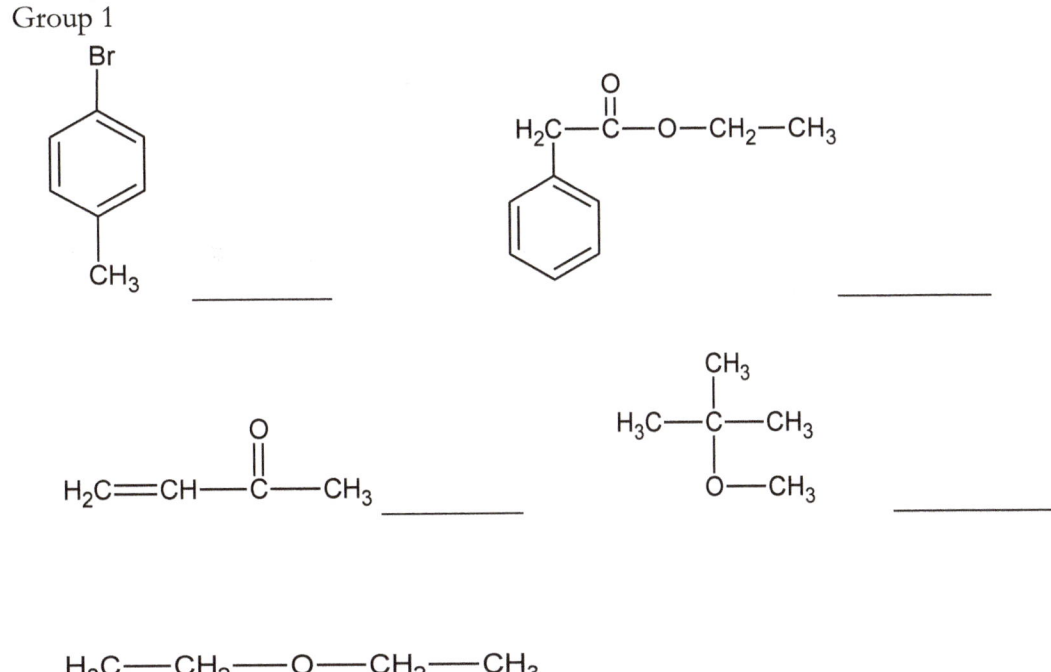

a.

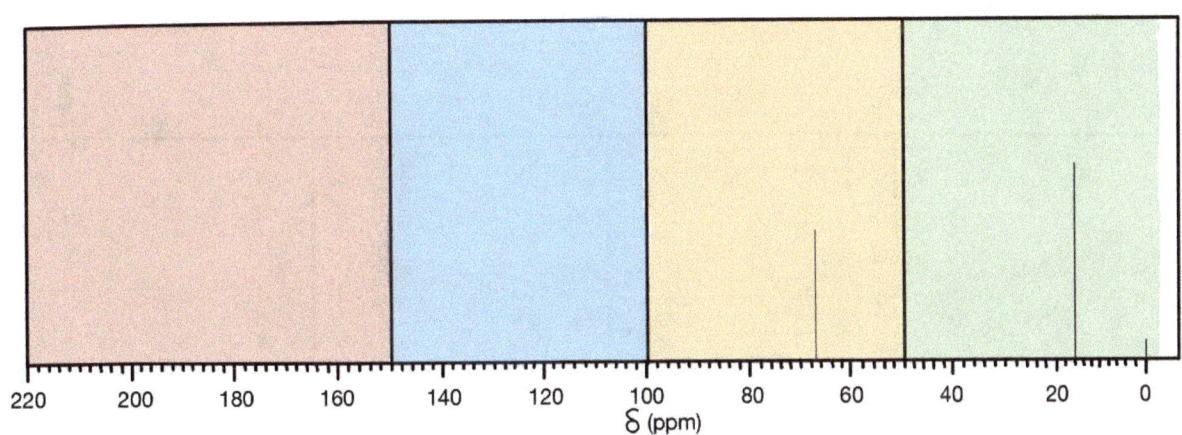

b.

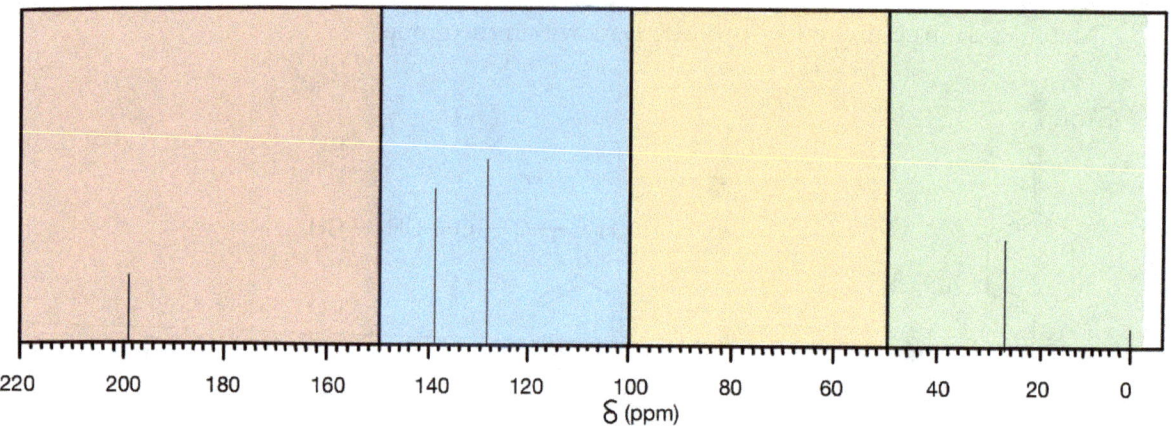

c.

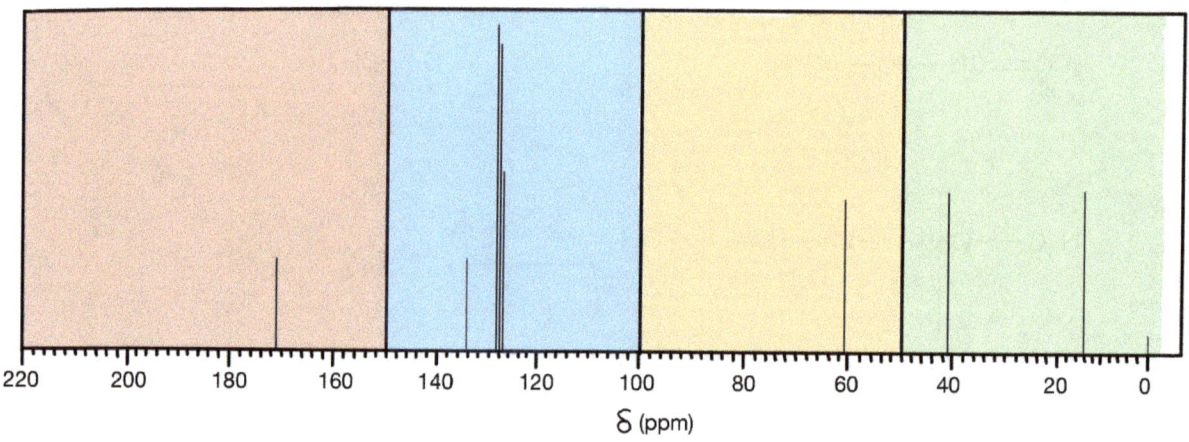

d.

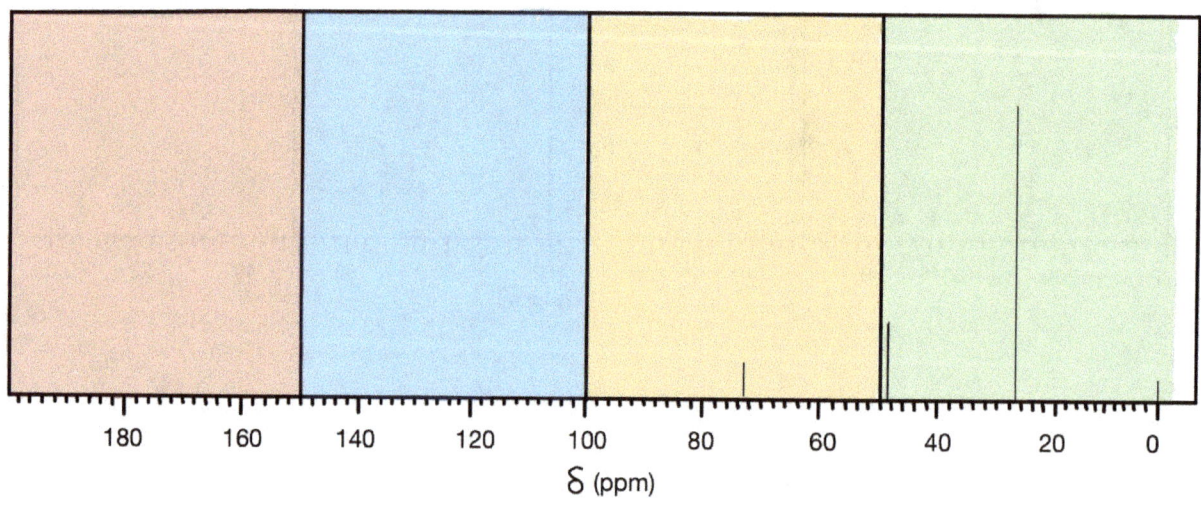

e.

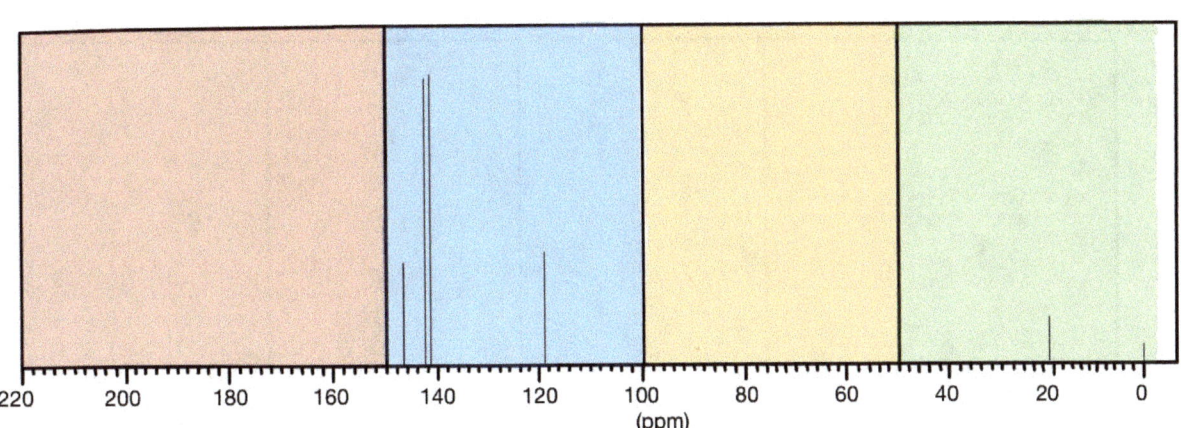

Group 2

H_3C—CH_2—$\overset{\displaystyle O}{\overset{\|}{C}}$—$CH_2$—$CH_3$ _____

H_3C—$\overset{\displaystyle CH_3}{\underset{\displaystyle CH_3}{\overset{\displaystyle |}{\underset{\displaystyle |}{C}}}}$—$Br$ _____

H_2C=CH—$\overset{\displaystyle O}{\overset{\|}{C}}$—$O$—$CH_3$ _____

H_3C—$\overset{\displaystyle CH_3}{\overset{\displaystyle |}{CH}}$—$CH_2$—$OH$ _____

H_3C—CH_2—$\overset{\displaystyle O}{\overset{\|}{C}}$—$O$—$\overset{\displaystyle CH_3}{\overset{\displaystyle |}{CH}}$—$CH_3$ _____

H_3C—$\overset{\displaystyle CH_3}{\overset{\displaystyle |}{CH}}$—$CH_2$—$\overset{\displaystyle O}{\overset{\|}{C}}$—$H$ _____

a.

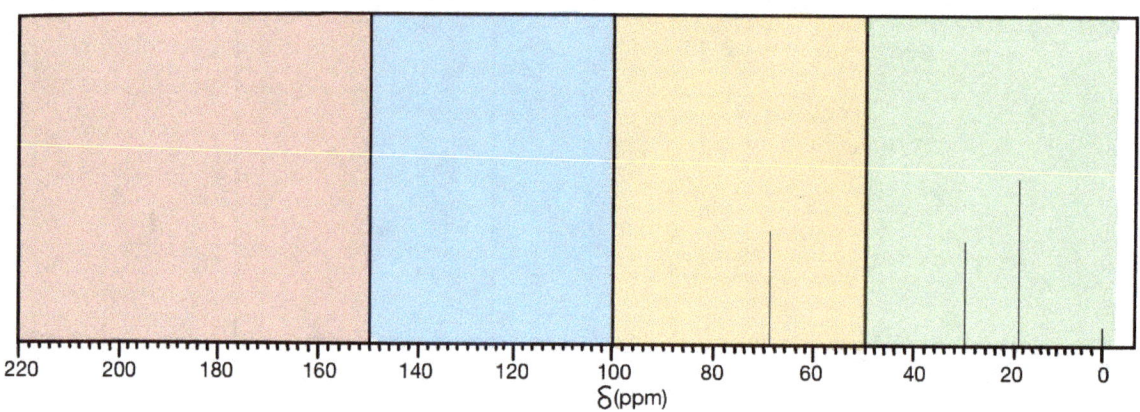

b.

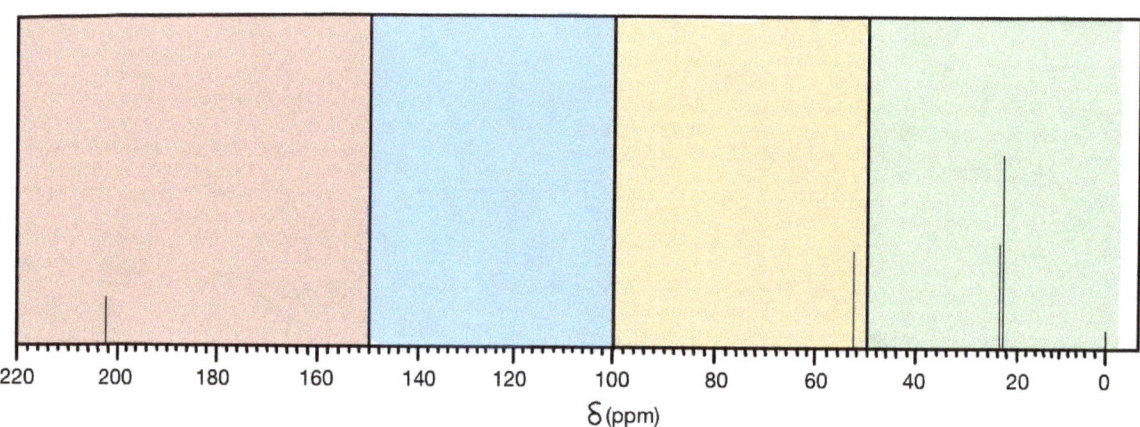

c.

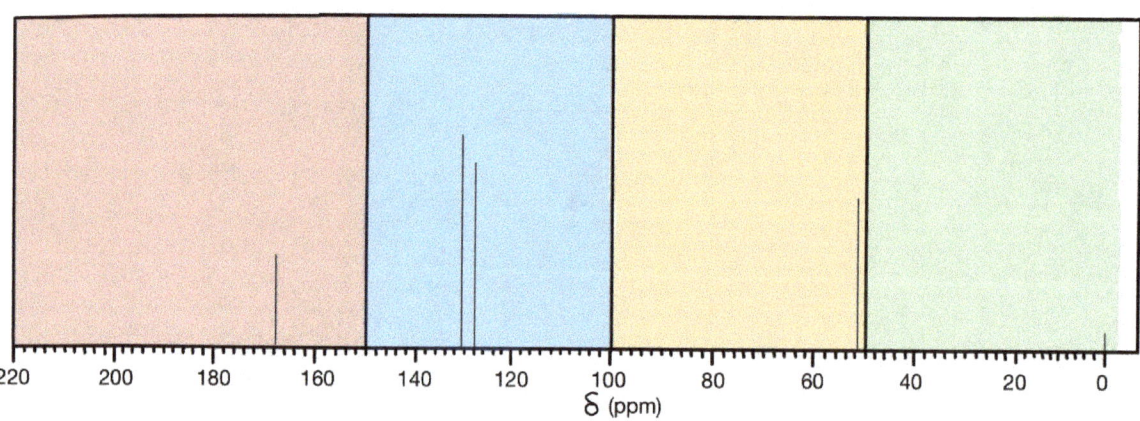

d.

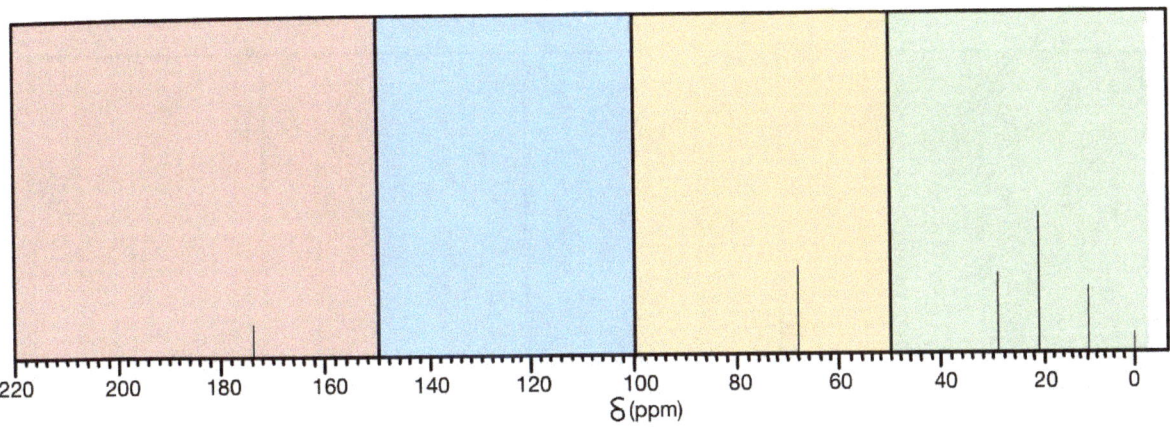

e.

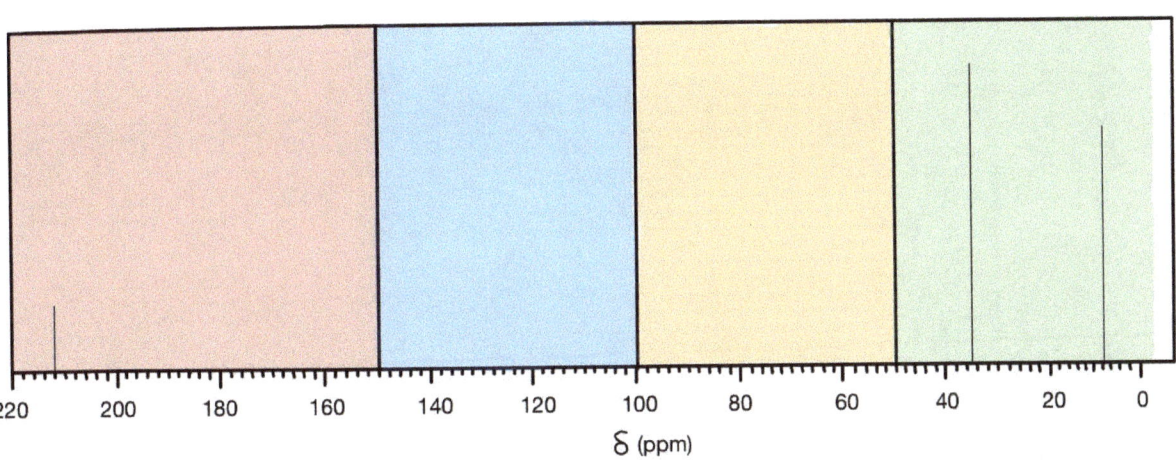

f.

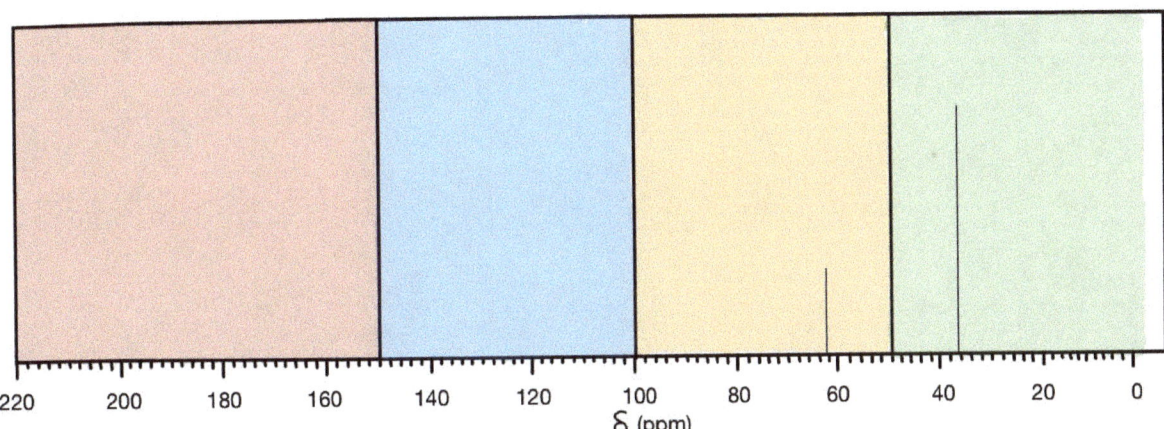

6. A substance with a molecular formula of C_6H_{12} produces the following coupled ^{13}C NMR spectrum. What is the structure of the molecule?

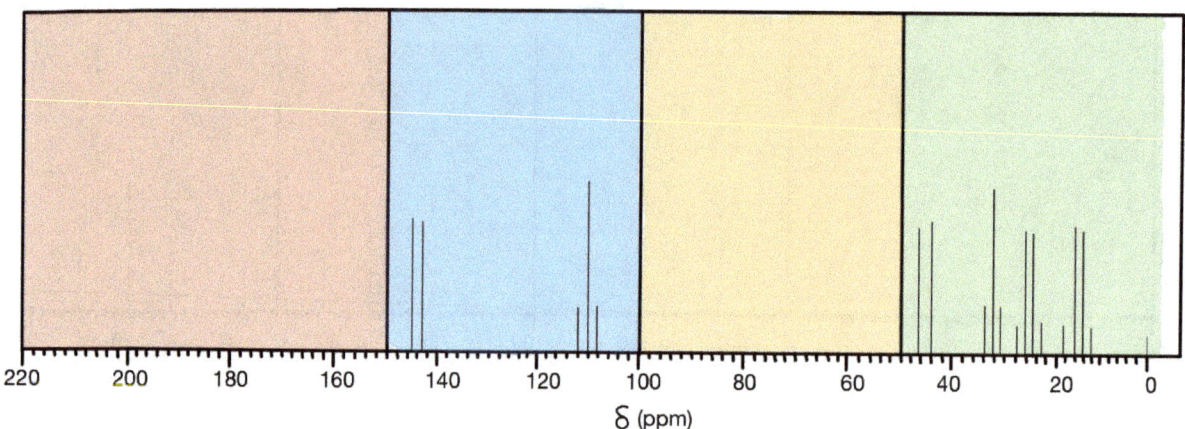

Mash-up 1

1. (Chapters 1 and 2) Write the IUPAC name for the following compound:

Stereochemistry

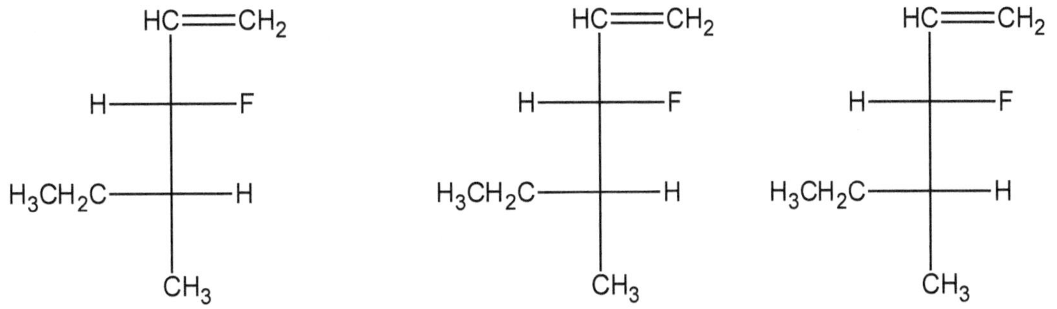

2. (Chapters 10 and 11) Write the IUPAC and common name for the following compound:

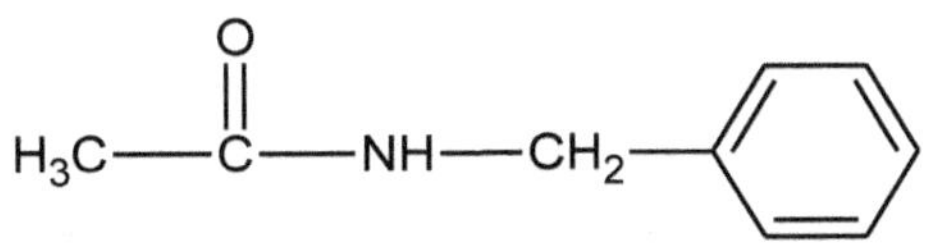

IUPAC name: _____

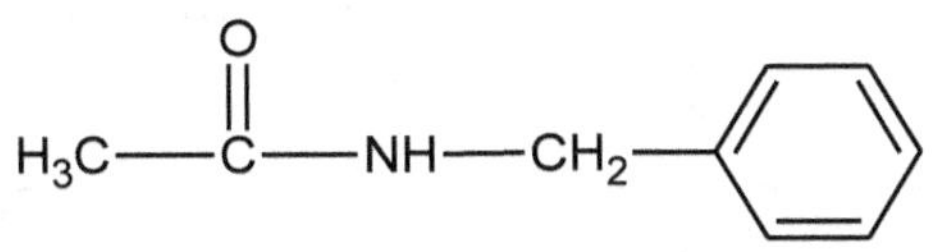

Common name: _____

3. (Chapter 10) Write the common name for the following compound:

H_2C—CH_3

C≡N

_____

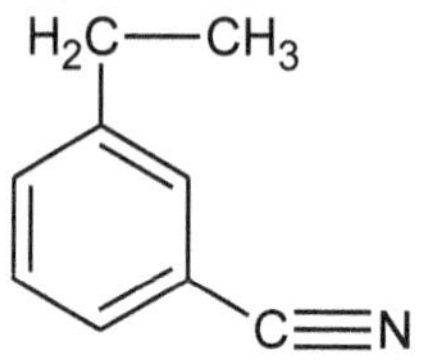

4. (Chapters 1 and 4) Write the pure IUPAC name for the following compound:

CH_3 CH_3

CH—CH—CH_3

_____

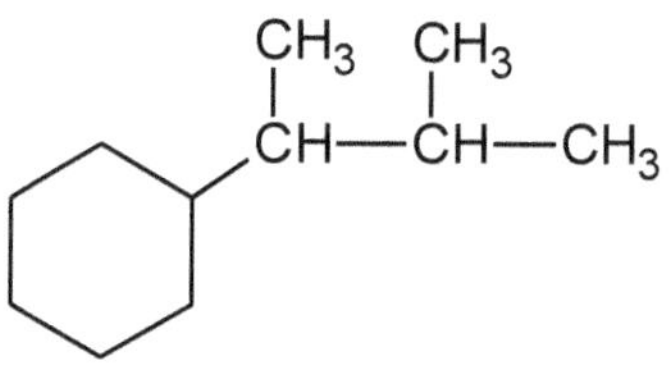

5. (Chapter 13) What type of instability will need to be addressed in the intermediate that is created when an sp^3-hybridized carbon that has a leaving group is reacted in a protic polar solvent?

_____

6. (Chapter 12) What type of instability will need to be addressed in the intermediate that is created when the carbon adjacent to a carbonyl carbon reacts with a nucleophile/base?

_____

7. (Chapter 9) What general methods could potentially be used to stabilize a carbanion?

_____

_____

_____

8. (Chapter 9) Which molecule in the following set is more stable and why?

$HC\!=\!CH_2$ $H_2C\!-\!CH_3$

_____

_____

9. (Chapters 15 and 16) If you needed to determine the identity of a proton NMR signal in the 6–8 ppm range, where would you look on an IR chart to distinguish whether it came from a vinyl hydrogen or an aryl hydrogen?

_____

_____

(Chapters 5–13) For each of the following,

10. a. Label the reactive features, highlight the most reactive feature, and then highlight what it needs. Also, state if a carbocation, carbon radical, or carbanion will start to develop, and/or if aromatic character will be lost because of a reaction between these molecules. If a carbocation, carbon radical or carbanion starts to develop, label where that will occur.

$+ HNO_3$ with tr. H_2SO_4

b. Use mechanism arrows to illustrate the reaction that occurs.

If applicable, use stabilization resources to deal with the carbocation, carbon radical, or carbanion that starts to develop during the reaction, and draw the structure of any resonance-stabilized intermediate.

Continue labelling and diagramming the reaction until you find the major stable product(s).

Finally, state the stereochemistry of the major product(s) and use either Fisher projection or perspective formula representations to illustrate that stereochemistry.

11. a. Label the reactive features, highlight the most reactive feature, and then highlight what it needs. Also, state if a carbocation, carbon radical, or carbanion will start to develop, and/or if aromatic character will be lost because of a reaction between these molecules. If a carbocation, carbon radical or carbanion starts to develop, label where that will occur.

(3S,4R)-4-methylhexan-3-ol + concentrated HCl

b. Use mechanism arrows to illustrate the reaction that occurs.

If applicable, use stabilization resources to deal with the carbocation, carbon radical, or carbanion that starts to develop during the reaction, and draw the structure of any resonance-stabilized intermediate.

Continue labelling and diagramming the reaction until you find the major stable product(s).

Finally, state the stereochemistry of the major product(s) and use either Fisher projection or perspective formula representations to illustrate that stereochemistry.

12. a. Label the reactive features, highlight the most reactive feature, and then highlight what it needs. Also, state if a carbocation, carbon radical, or carbanion will start to develop, and/or if aromatic character will be lost because of a reaction between these molecules. If a carbocation, carbon radical or carbanion starts to develop, label where that will occur.

Acetone in methanol with catalytic amounts of acid

b. Use mechanism arrows to illustrate the reaction that occurs.

If applicable, use stabilization resources to deal with the carbocation, carbon radical, or carbanion that starts to develop during the reaction, and draw the structure of any resonance-stabilized intermediate.

Continue labelling and diagramming the reaction until you find the major stable product(s).

Finally, state the stereochemistry of the major product(s) and use either Fisher projection or perspective formula representations to illustrate that stereochemistry.

13. a. Label the reactive features, highlight the most reactive feature, and then highlight what it needs. Also, state if a carbocation, carbon radical, or carbanion will start to develop, and/or if aromatic character will be lost because of a reaction between these molecules. If a carbocation, carbon radical or carbanion starts to develop, label where that will occur.

m-isopropylphenol + HNO₃ with tr. H₂SO₄

b. Use mechanism arrows to illustrate the reaction that occurs.

If applicable, use stabilization resources to deal with the carbocation, carbon radical, or carbanion that starts to develop during the reaction, and draw the structure of any resonance-stabilized intermediate.

Continue labelling and diagramming the reaction until you find the major stable product(s).

Finally, state the stereochemistry of the major product(s) and use either Fisher projection or perspective formula representations to illustrate that stereochemistry.

14. a. Label the reactive features, highlight the most reactive feature, and then highlight what it needs. Also, state if a carbocation, carbon radical, or carbanion will start to develop, and/or if aromatic character will be lost because of a reaction between these molecules. If a carbocation, carbon radical or carbanion starts to develop, label where that will occur.

nitrobenzene + 2-chloro-3-methyl butane with $AlCl_3$

b. Use mechanism arrows to illustrate the reaction that occurs.

If applicable, use stabilization resources to deal with the carbocation, carbon radical, or carbanion that starts to develop during the reaction, and draw the structure of any resonance-stabilized intermediate.

Continue labelling and diagramming the reaction until you find the major stable product(s).

Finally, state the stereochemistry of the major product(s) and use either Fisher projection or perspective formula representations to illustrate that stereochemistry.

15. a. Label the reactive features, highlight the most reactive feature, and then highlight what it needs. Also, state if a carbocation, carbon radical, or carbanion will start to develop, and/or if aromatic character will be lost because of a reaction between these molecules. If a carbocation, carbon radical or carbanion starts to develop, label where that will occur.

1,2-dimethylcyclopentene in ethanol with a tr. of H_2SO_4

b. Use mechanism arrows to illustrate the reaction that occurs.

If applicable, use stabilization resources to deal with the carbocation, carbon radical, or carbanion that starts to develop during the reaction, and draw the structure of any resonance-stabilized intermediate.

Continue labelling and diagramming the reaction until you find the major stable product(s).

Finally, state the stereochemistry of the major product(s) and use either Fisher projection or perspective formula representations to illustrate that stereochemistry.

16. a. Label the reactive features, highlight the most reactive feature, and then highlight what it needs. Also, state if a carbocation, carbon radical, or carbanion will start to develop, and/or if aromatic character will be lost because of a reaction between these molecules. If a carbocation, carbon radical or carbanion starts to develop, label where that will occur.

+ NaOH

b. Use mechanism arrows to illustrate the reaction that occurs.

If applicable, use stabilization resources to deal with the carbocation, carbon radical, or carbanion that starts to develop during the reaction, and draw the structure of any resonance-stabilized intermediate.

Continue labelling and diagramming the reaction until you find the major stable product(s).

Finally, state the stereochemistry of the major product(s) and use either Fisher projection or perspective formula representations to illustrate that stereochemistry.

17. a. Label the reactive features, highlight the most reactive feature, and then highlight what it needs. Also, state if a carbocation, carbon radical, or carbanion will start to develop, and/or if aromatic character will be lost because of a reaction between these molecules. If a carbocation, carbon radical or carbanion starts to develop, label where that will occur.

2-methylcyclohexa-1,3-diene with HBr

b. Use mechanism arrows to illustrate the reaction that occurs.

If applicable, use stabilization resources to deal with the carbocation, carbon radical, or carbanion that starts to develop during the reaction, and draw the structure of any resonance-stabilized intermediate.

Continue labelling and diagramming the reaction until you find the major stable product(s).

Finally, state the stereochemistry of the major product(s) and use either Fisher projection or perspective formula representations to illustrate that stereochemistry.

18. (Chapters 1, 13, 15, and 16) Analysis of a compound with the molecular formula of $C_7H_{16}O$ produced the following IR and NMR data:

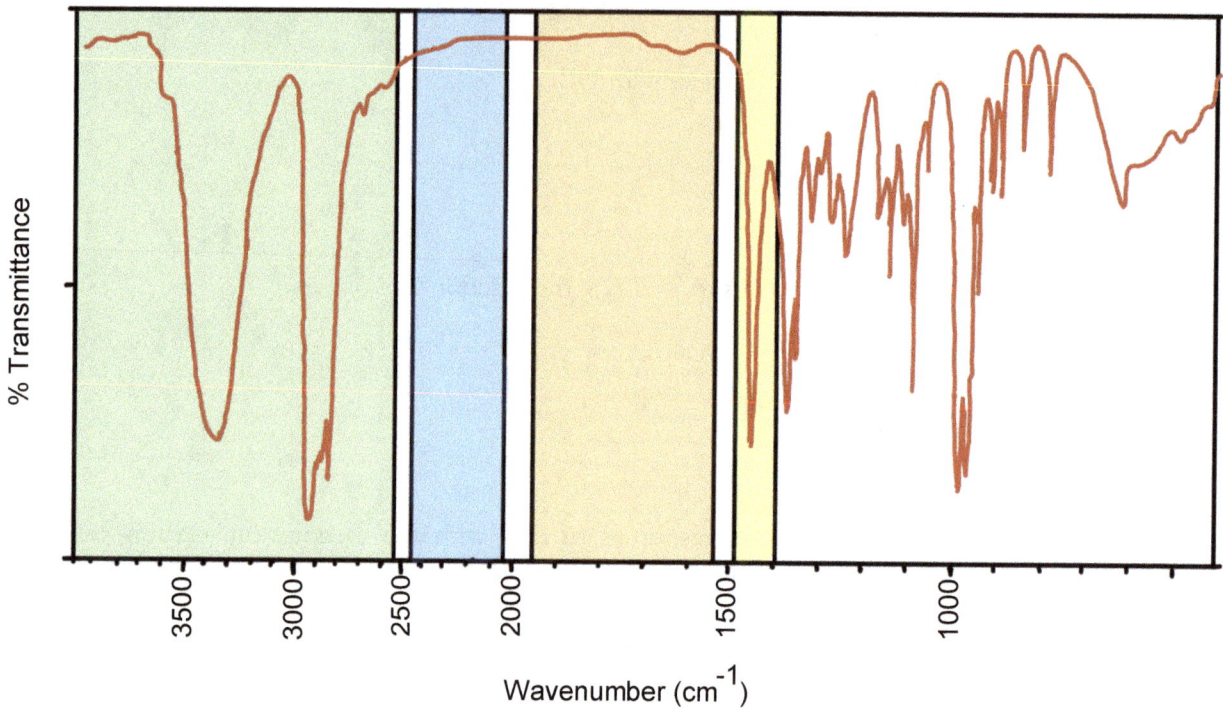

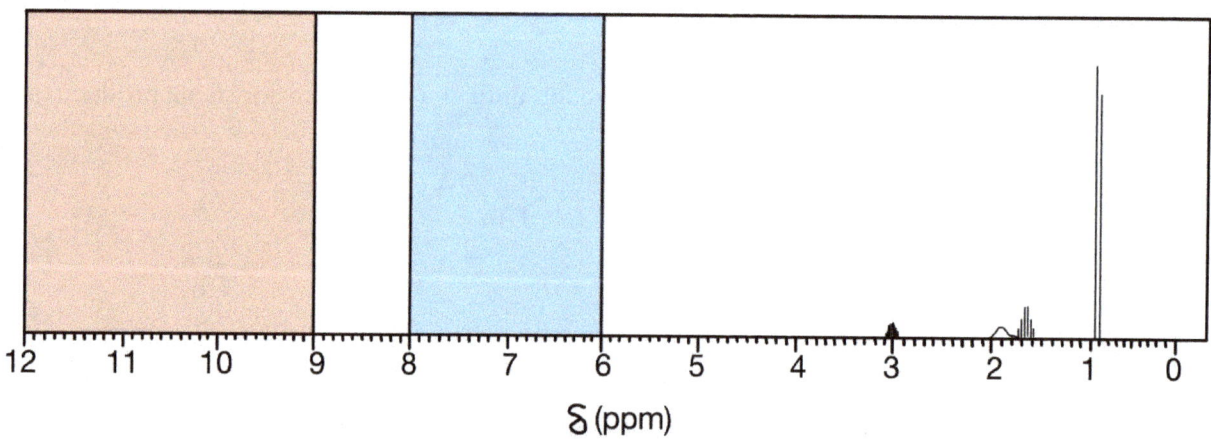

a. Draw the structure of the compound.

b. Write the IUPAC name for the compound:

_____

c. Propose the best method for synthesizing this compound from an alkyl halide.

_____

19. (Chapters 1, 3, 6, 14, 15, and 16) MS, IR, and NMR analysis of a compound produced the following data:

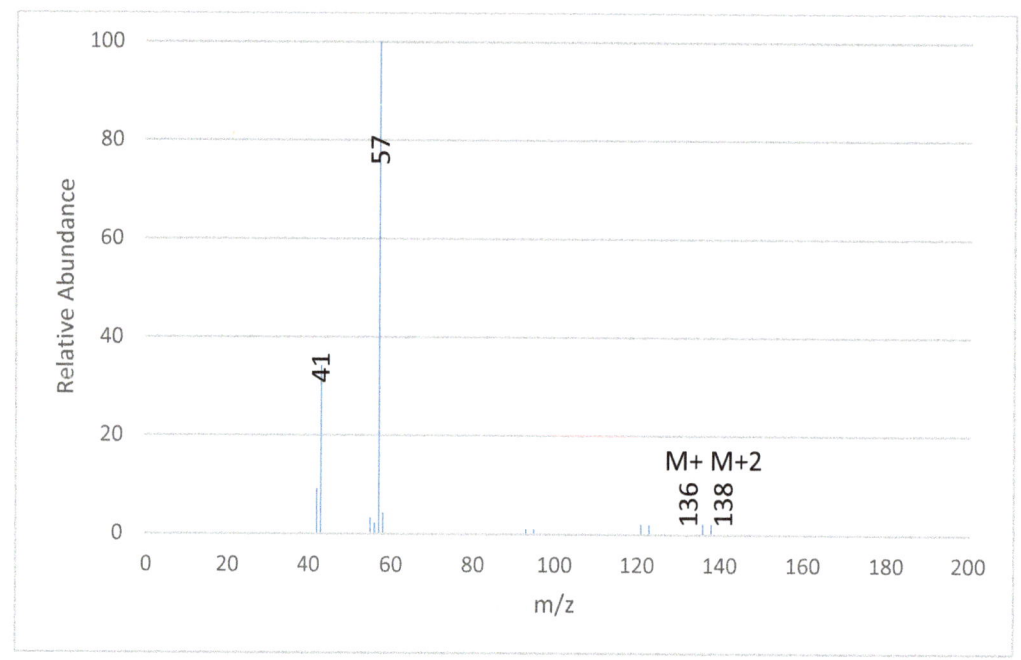

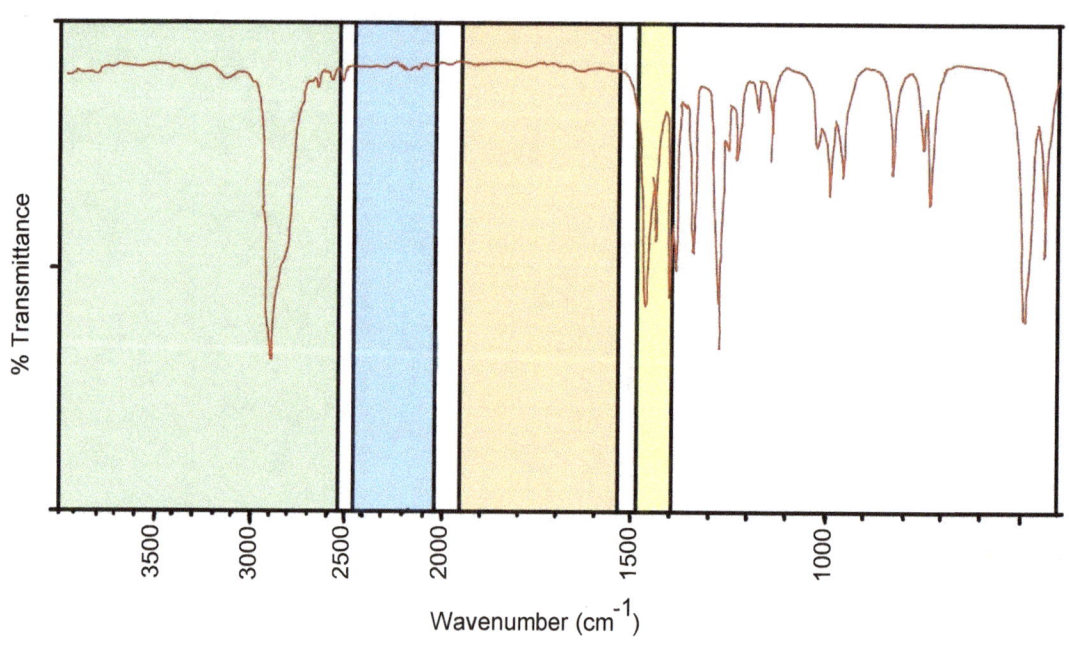

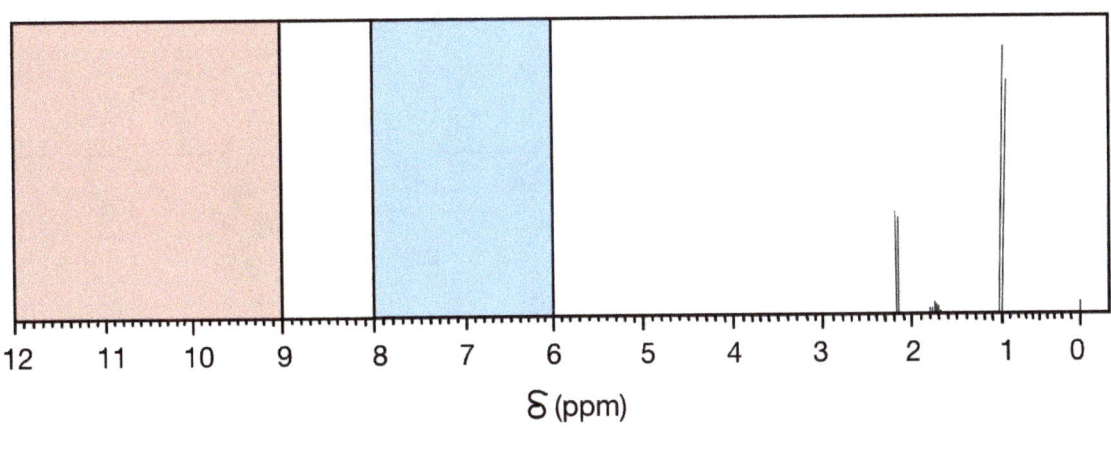

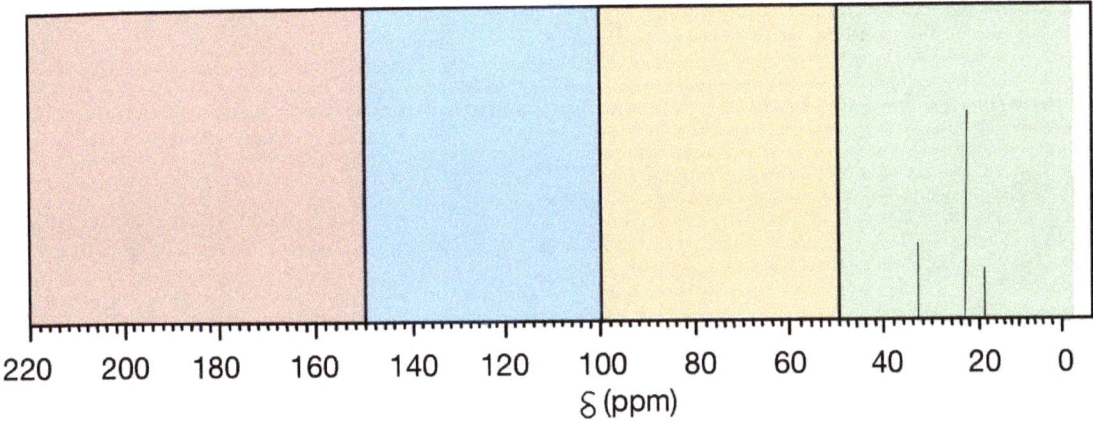

a. Draw the structure of the compound.

b. Write the IUPAC and the common name for the compound.

IUPAC: _____

Common name: _____

c. Propose the best method for synthesizing this compound starting with any alkene.

_____

Mash-up 2

1. (Chapters 1 and 2) Write the IUPAC name for the following compound:

Stereochemistry

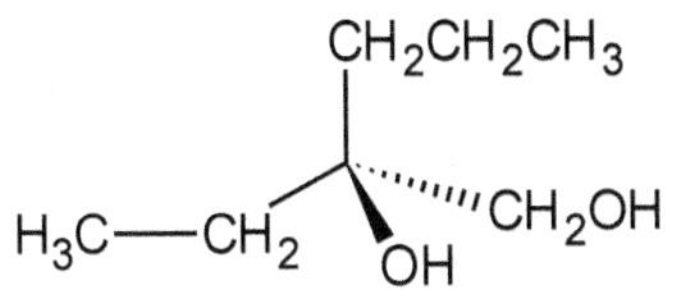

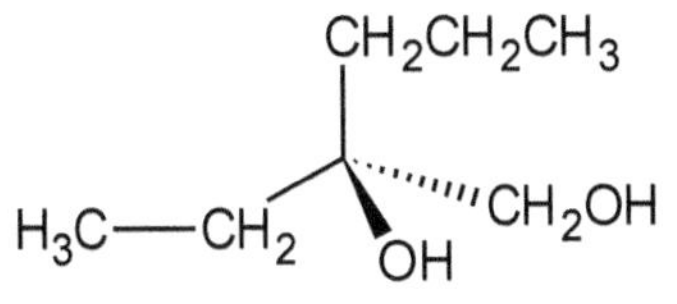

2. (Chapters 1 and 3) Write the IUPAC and common name for the following molecule:

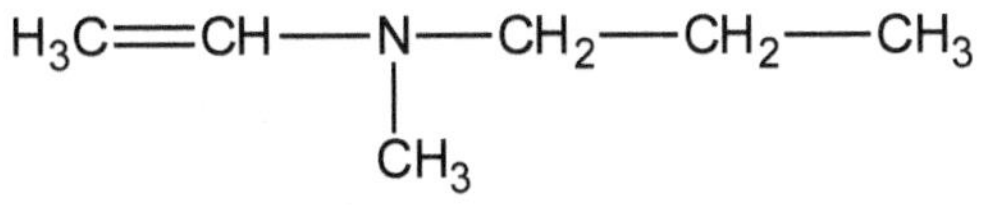

IUPAC: _____

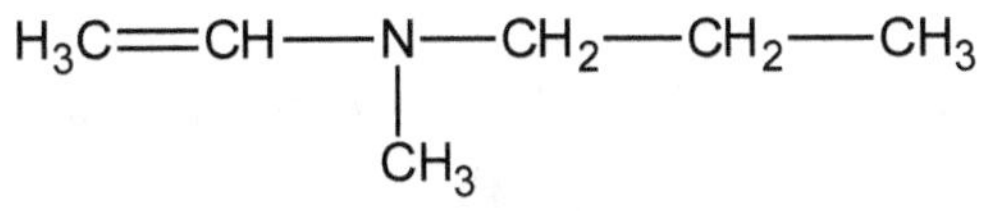

Common name: _____

3. (Chapter 11) Write the IUPAC and common name for the following compound:

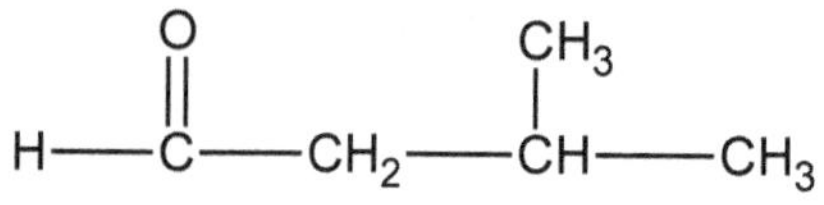

$$H-\overset{\overset{\displaystyle O}{\|}}{C}-CH_2-\overset{\overset{\displaystyle CH_3}{|}}{CH}-CH_3$$

IUPAC: _____

$$H-\overset{\overset{\displaystyle O}{\|}}{C}-CH_2-\overset{\overset{\displaystyle CH_3}{|}}{CH}-CH_3$$

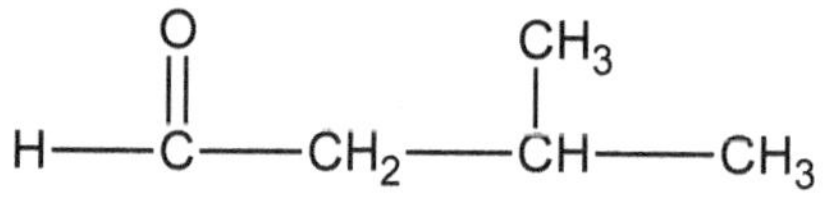

Common name: _____

4. (Chapter 10) Write the common name for the following compound:

$$\overset{\displaystyle H_3C\diagdown_{CH}\diagup CH_3}{}$$

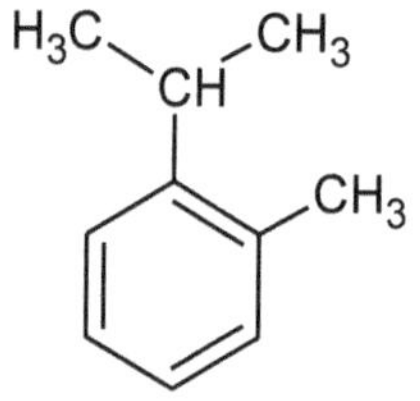

5. (Chapter 6) What type of instability will need to be addressed in the intermediate that is created when a pi bond reacts with a hydrogen that has a partial positive charge?

_____

6. (Chapter 10) What type of instability will need to be addressed in the intermediate that is created when an aromatic compound reacts with a very reactive electrophile?

_____

7. (Chapter 13) What general resources could potentially be used to react with a carbon that has a partial positive charge?

_____

_____

_____

8. (Chapter 9) Which molecule in the following set is more stable and why?

$$H_2C-\overset{\overset{\displaystyle O}{\|}}{C}-CH_2-CH_3 \qquad H_2C-\overset{\overset{\displaystyle NH}{\|}}{C}-CH_2-CH_3$$

_____

_____

9. (Chapters 15 and 16) If you saw an NMR signal in the 10–12 range, at what approximate IR wavenumbers would you look to distinguish between the two possibilities?

_____

_____

(Chapters 5–13) For each of the following:
10. a. Label the reactive features, highlight the most reactive feature, and then highlight what it needs. Also, state if a carbocation, carbon radical, or carbanion will start to develop, and/or if aromatic character will be lost because of a reaction between these molecules. If a carbocation, carbon radical or carbanion starts to develop, label where that will occur.

benzenenitrile + Cl_2 and $FeCl_3$

b. Use mechanism arrows to illustrate the reaction that occurs.

If applicable, use stabilization resources to deal with the carbocation, carbon radical, or carbanion that starts to develop during the reaction, and draw the structure of any resonance-stabilized intermediate.

Continue labelling and diagramming the reaction until you find the major stable product(s).

Finally, state the stereochemistry of the major product(s) and use either Fisher projection or perspective formula representations to illustrate that stereochemistry.

11. a. Label the reactive features, highlight the most reactive feature, and then highlight what it needs. Also, state if a carbocation, carbon radical, or carbanion will start to develop, and/or if aromatic character will be lost because of a reaction between these molecules. If a carbocation, carbon radical or carbanion starts to develop, label where that will occur.

(3S,4R)-3-bromo-4-methylhexane in methanol

b. Use mechanism arrows to illustrate the reaction that occurs.

If applicable, use stabilization resources to deal with the carbocation, carbon radical, or carbanion that starts to develop during the reaction, and draw the structure of any resonance-stabilized intermediate.

Continue labelling and diagramming the reaction until you find the major stable product(s).

Finally, state the stereochemistry of the major product(s) and use either Fisher projection or perspective formula representations to illustrate that stereochemistry.

12. a. Label the reactive features, highlight the most reactive feature, and then highlight what it needs. Also, state if a carbocation, carbon radical, or carbanion will start to develop, and/or if aromatic character will be lost because of a reaction between these molecules. If a carbocation, carbon radical or carbanion starts to develop, label where that will occur.

(R)-3-methylpent-1-ene with $Hg(OAc)_2$ in ethanol, followed by $NaBH_4$

b. Use mechanism arrows to illustrate the reaction that occurs.

If applicable, use stabilization resources to deal with the carbocation, carbon radical, or carbanion that starts to develop during the reaction, and draw the structure of any resonance-stabilized intermediate.

Continue labelling and diagramming the reaction until you find the major stable product(s).

Finally, state the stereochemistry of the major product(s) and use either Fisher projection or perspective formula representations to illustrate that stereochemistry.

13. a. Label the reactive features, highlight the most reactive feature, and then highlight what it needs. Also, state if a carbocation, carbon radical, or carbanion will start to develop, and/or if aromatic character will be lost because of a reaction between these molecules. If a carbocation, carbon radical or carbanion starts to develop, label where that will occur.

1-methylcyclohexa-1,3-diene with Cl_2

b. Use mechanism arrows to illustrate the reaction that occurs.

If applicable, use stabilization resources to deal with the carbocation, carbon radical, or carbanion that starts to develop during the reaction, and draw the structure of any resonance-stabilized intermediate.

Continue labelling and diagramming the reaction until you find the major stable product(s).

Finally, state the stereochemistry of the major product(s) and use either Fisher projection or perspective formula representations to illustrate that stereochemistry.

14. a. Label the reactive features, highlight the most reactive feature, and then highlight what it needs. Also, state if a carbocation, carbon radical, or carbanion will start to develop, and/or if aromatic character will be lost because of a reaction between these molecules. If a carbocation, carbon radical or carbanion starts to develop, label where that will occur.

o-propylbenzaldehyde + Cl$_2$ and FeCl$_3$

b. Use mechanism arrows to illustrate the reaction that occurs.

If applicable, use stabilization resources to deal with the carbocation, carbon radical, or carbanion that starts to develop during the reaction, and draw the structure of any resonance-stabilized intermediate.

Continue labelling and diagramming the reaction until you find the major stable product(s).

Finally, state the stereochemistry of the major product(s) and use either Fisher projection or perspective formula representations to illustrate that stereochemistry.

15. a. Label the reactive features, highlight the most reactive feature, and then highlight what it needs. Also, state if a carbocation, carbon radical, or carbanion will start to develop, and/or if aromatic character will be lost because of a reaction between these molecules. If a carbocation, carbon radical or carbanion starts to develop, label where that will occur.

   (3S,4R)-4-methylhexan-3-ol + POCl₃ in pyridine

   b. Use mechanism arrows to illustrate the reaction that occurs.

   If applicable, use stabilization resources to deal with the carbocation, carbon radical, or carbanion that starts to develop during the reaction, and draw the structure of any resonance-stabilized intermediate.

   Continue labelling and diagramming the reaction until you find the major stable product(s).

   Finally, state the stereochemistry of the major product(s) and use either Fisher projection or perspective formula representations to illustrate that stereochemistry.

16. a. Label the reactive features, highlight the most reactive feature, and then highlight what it needs. Also, state if a carbocation, carbon radical, or carbanion will start to develop, and/or if aromatic character will be lost because of a reaction between these molecules. If a carbocation, carbon radical or carbanion starts to develop, label where that will occur.

cyclopentan-2-one in NH_3 with catalytic amounts of acid

b. Use mechanism arrows to illustrate the reaction that occurs.

If applicable, use stabilization resources to deal with the carbocation, carbon radical, or carbanion that starts to develop during the reaction, and draw the structure of any resonance-stabilized intermediate.

Continue labelling and diagramming the reaction until you find the major stable product(s).

Finally, state the stereochemistry of the major product(s) and use either Fisher projection or perspective formula representations to illustrate that stereochemistry.

17. a. Label the reactive features, highlight the most reactive feature, and then highlight what it needs. Also, state if a carbocation, carbon radical, or carbanion will start to develop, and/or if aromatic character will be lost because of a reaction between these molecules. If a carbocation, carbon radical or carbanion starts to develop, label where that will occur.

$$CH_3$$
$$HC-CH_3$$

NH—CH$_3$ + 2-methylpropanoyl chloride + AlCl$_3$

b. Use mechanism arrows to illustrate the reaction that occurs.

If applicable, use stabilization resources to deal with the carbocation, carbon radical, or carbanion that starts to develop during the reaction, and draw the structure of any resonance-stabilized intermediate.

Continue labelling and diagramming the reaction until you find the major stable product(s).

Finally, state the stereochemistry of the major product(s) and use either Fisher projection or perspective formula representations to illustrate that stereochemistry.

18. a. Label the reactive features, highlight the most reactive feature, and then highlight what it needs. Also, state if a carbocation, carbon radical, or carbanion will start to develop, and/or if aromatic character will be lost because of a reaction between these molecules. If a carbocation, carbon radical or carbanion starts to develop, label where that will occur.

   3-methylcyclohexene + HBr

   b. Use mechanism arrows to illustrate the reaction that occurs.

   If applicable, use stabilization resources to deal with the carbocation, carbon radical, or carbanion that starts to develop during the reaction, and draw the structure of any resonance-stabilized intermediate.

   Continue labelling and diagramming the reaction until you find the major stable product(s).

   Finally, state the stereochemistry of the major product(s) and use either Fisher projection or perspective formula representations to illustrate that stereochemistry.

19. (Chapters 1, 13, 15, and 16) A compound with the molecular formula of $C_6H_{14}O$ produced the following IR and NMR data:

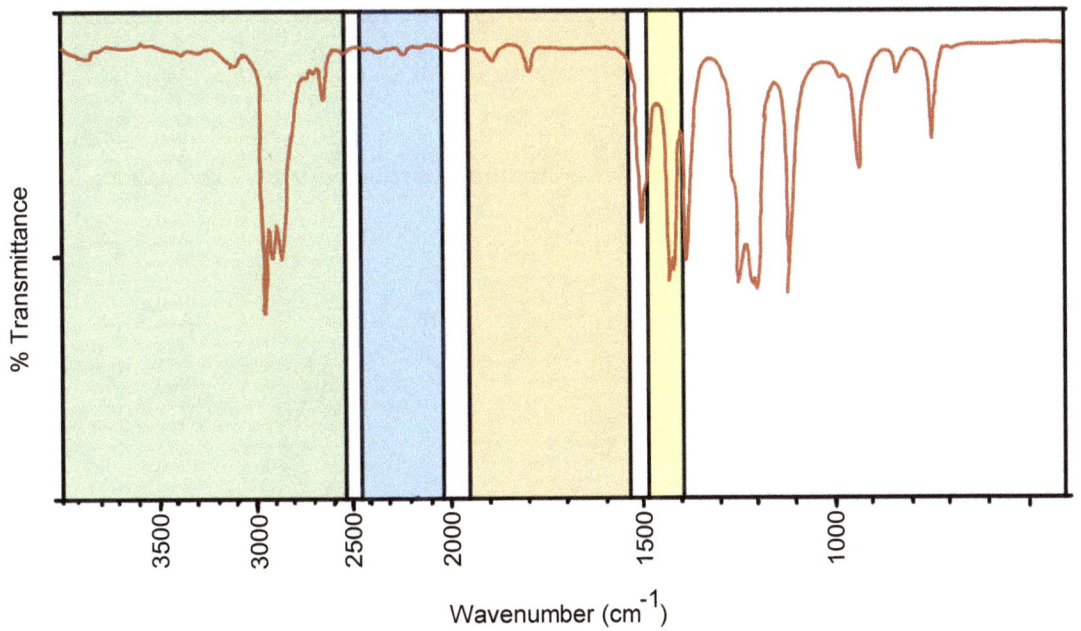

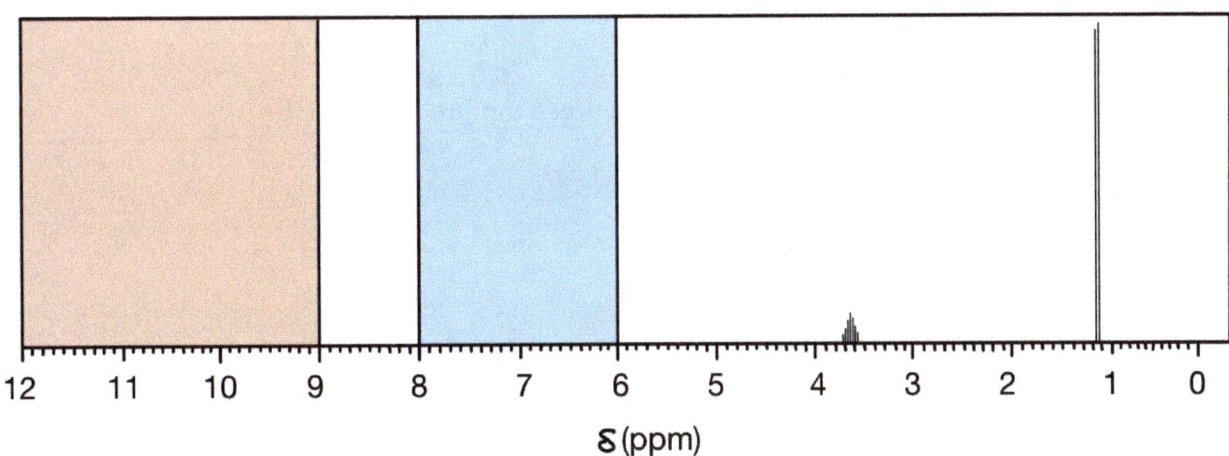

a. Draw the structure of the compound.

b. Write the IUPAC name for the compound.

_____

c. Propose a method for synthesizing this compound starting with any alkyl halide.

d. Predict the major product of a reaction between this product and HBr.

20. (Chapters 1, 3, 6, 14, 15, and 16) MS, IR, and NMR analysis of a compound produced the following data:

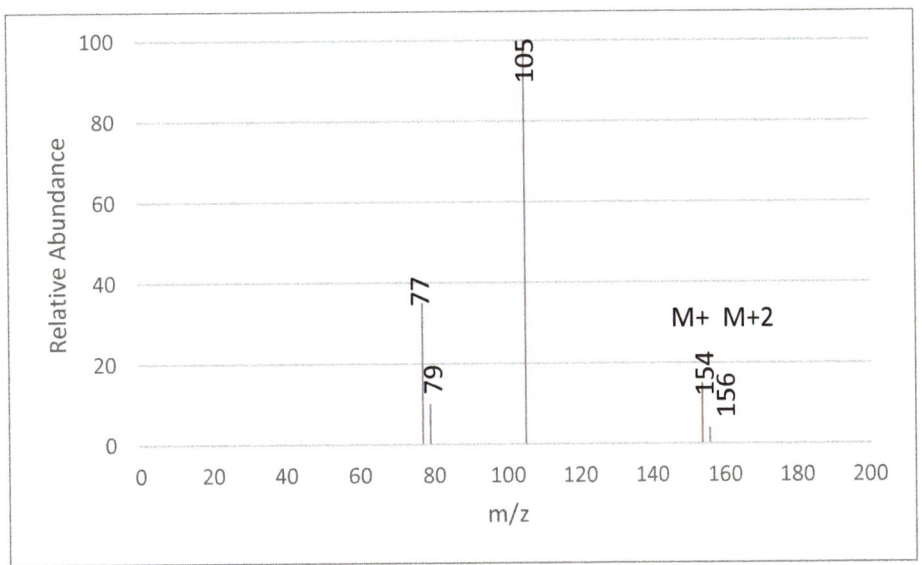

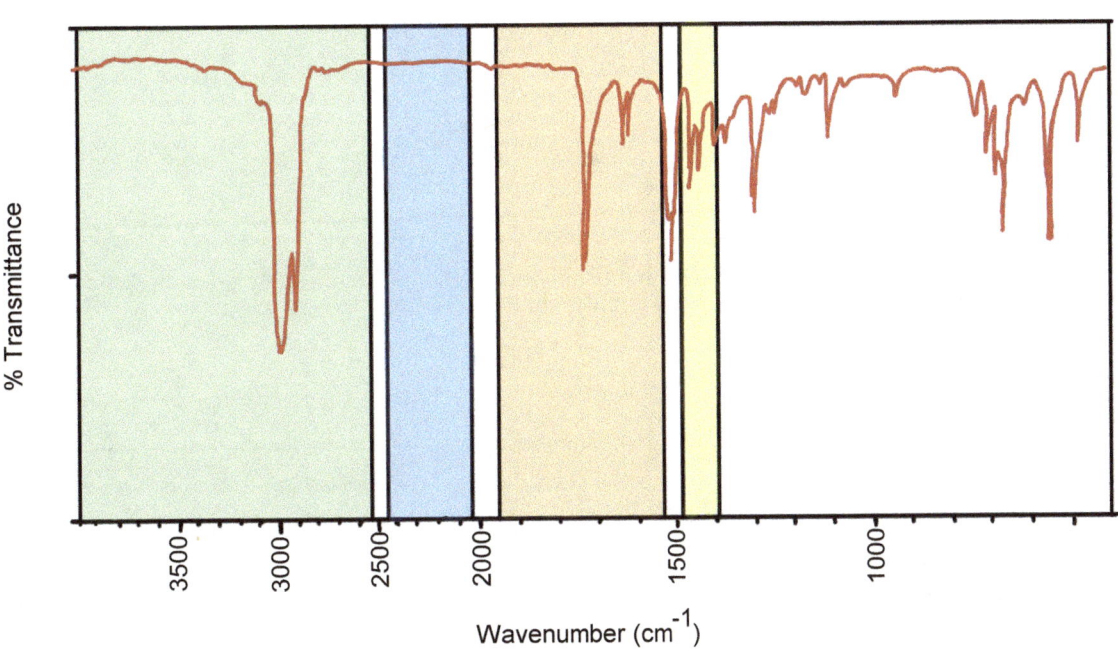

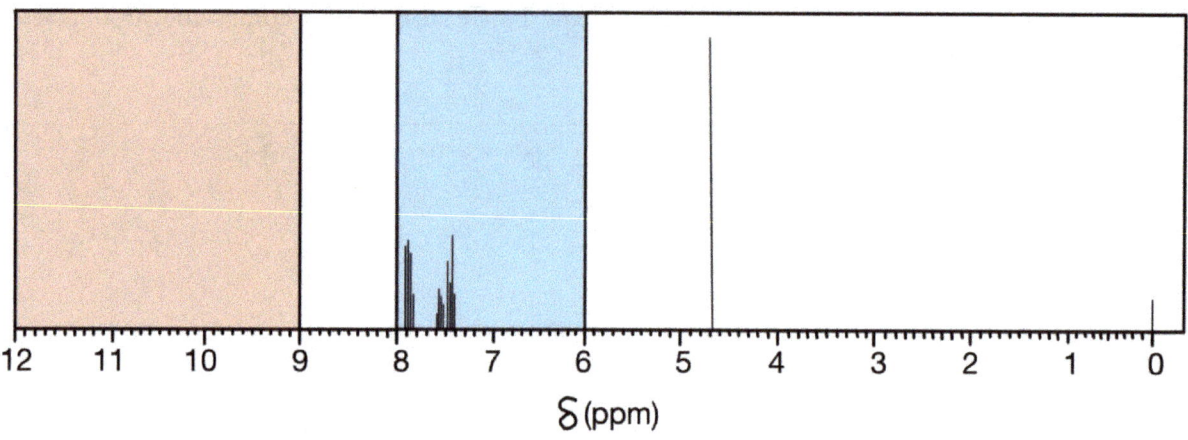

a. Draw the structure of the compound.

_____

b. Write the IUPAC name for the compound.

_____

c. Propose a method for synthesizing this compound starting with any alcohol and any other reagents.

_____

Mash-up 3

1. (Chapters 1 and 3) Write the IUPAC and common name for the following compound:

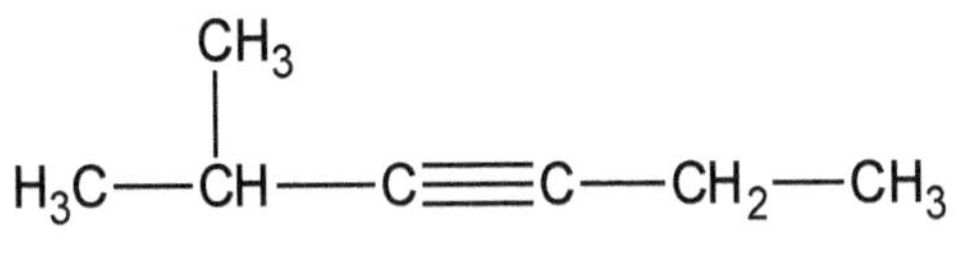

$$H_3C-\underset{\underset{\displaystyle CH_3}{|}}{CH}-C\equiv C-CH_2-CH_3$$

IUPAC: _____

$$H_3C-\underset{\underset{\displaystyle CH_3}{|}}{CH}-C\equiv C-CH_2-CH_3$$

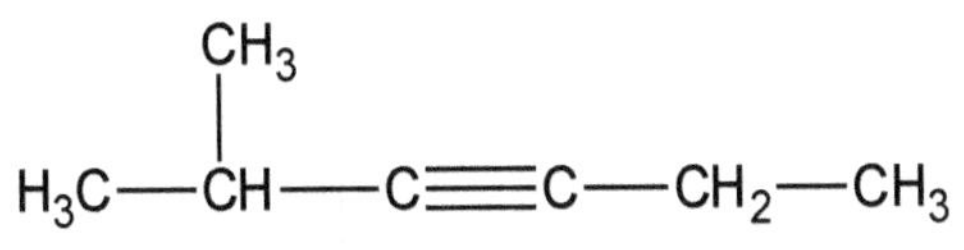

Common name: _____

2. (Chapters 1 and 2) Write the IUPAC name for the following compound:

$$H_3C-CH_2-CH_2-O-\underset{\displaystyle H_3C-CH_2}{\overset{}{C}}=C-\underset{\displaystyle H}{\overset{\displaystyle CH-CH_3}{\underset{\displaystyle |}{\overset{\displaystyle CH_3}{|}}}}$$

_____

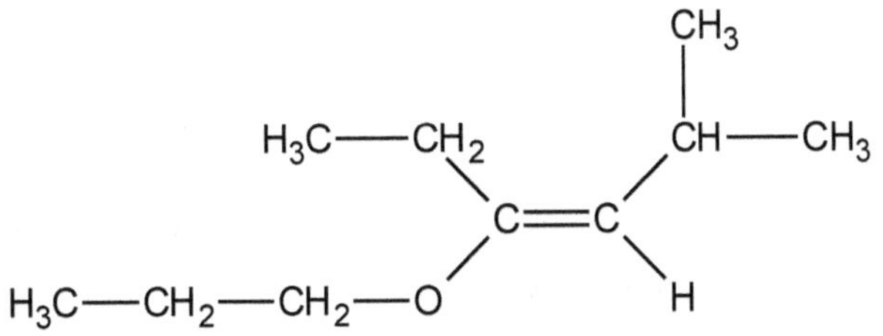

3. (Chapter 10) Write the common name for the following compound:

_____

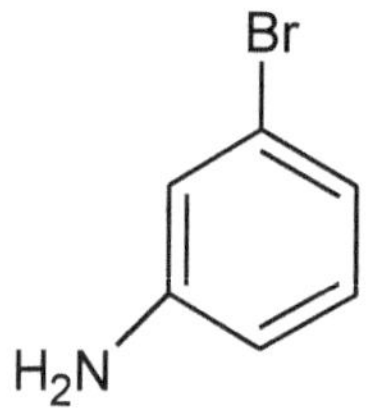

4. (Chapter 11) Write the IUPAC and common name for the following compound:

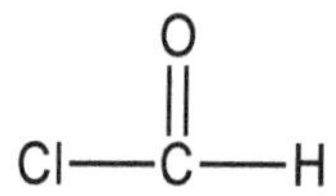

IUPAC: _____

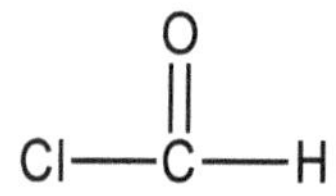

Common name: _____

5. (Chapters 15 and 16) If you saw a very shallow, narrow band at approximately 1650 nm^{-1} on an IR chart, where would you look on an NMR spectrum to determine whether or not the band is a significant factor to use when trying to derive the structure of the molecule?

_____

_____

6. (Chapters 5 and 6) What type of instability will need to be addressed in the intermediate that is created when a radical reacts with bond electrons?

_____

7. (Chapter 7) What type of instability will need to be addressed in the intermediate that is created when a pi bond reacts with an electrophilic atom that has a non-bonded electron pair?

_____

8. (Chapters 6 and 8) What general methods could potentially be used to stabilize a carbon radical?

_____

_____

_____

9. (Chapter 6) Which molecule in the following set is more stable and why?

_____

_____

(Chapters 5–13) For each of the following:

10. a. Label the reactive features, highlight the most reactive feature, and then highlight what it needs. Also, state if a carbocation, carbon radical, or carbanion will start to develop, and/or if aromatic character will be lost because of a reaction between these molecules. If a carbocation, carbon radical or carbanion starts to develop, label where that will occur.

3,3-dimethylcyclohexene + BH$_3$ followed by H$_2$O$_2$, $^-$OH, and H$_2$O

b. Use mechanism arrows to illustrate the reaction that occurs.

If applicable, use stabilization resources to deal with the carbocation, carbon radical, or carbanion that starts to develop during the reaction, and draw the structure of any resonance-stabilized intermediate.

Continue labelling and diagramming the reaction until you find the major stable product(s).

Finally, state the stereochemistry of the major product(s) and use either Fisher projection or perspective formula representations to illustrate that stereochemistry.

11. a. Label the reactive features, highlight the most reactive feature, and then highlight what it needs. Also, state if a carbocation, carbon radical, or carbanion will start to develop, and/or if aromatic character will be lost because of a reaction between these molecules. If a carbocation, carbon radical or carbanion starts to develop, label where that will occur.

isopentyl bromide with sodium methoxide

b. Use mechanism arrows to illustrate the reaction that occurs.

If applicable, use stabilization resources to deal with the carbocation, carbon radical, or carbanion that starts to develop during the reaction, and draw the structure of any resonance-stabilized intermediate.

Continue labelling and diagramming the reaction until you find the major stable product(s).

Finally, state the stereochemistry of the major product(s) and use either Fisher projection or perspective formula representations to illustrate that stereochemistry.

12. a. Label the reactive features, highlight the most reactive feature, and then highlight what it needs. Also, state if a carbocation, carbon radical, or carbanion will start to develop, and/or if aromatic character will be lost because of a reaction between these molecules. If a carbocation, carbon radical or carbanion starts to develop, label where that will occur.

p-isopropylaniline + H_2SO_4

b. Use mechanism arrows to illustrate the reaction that occurs.

If applicable, use stabilization resources to deal with the carbocation, carbon radical, or carbanion that starts to develop during the reaction, and draw the structure of any resonance-stabilized intermediate.

Continue labelling and diagramming the reaction until you find the major stable product(s).

Finally, state the stereochemistry of the major product(s) and use either Fisher projection or perspective formula representations to illustrate that stereochemistry.

13. a. Label the reactive features, highlight the most reactive feature, and then highlight what it needs. Also, state if a carbocation, carbon radical, or carbanion will start to develop, and/or if aromatic character will be lost because of a reaction between these molecules. If a carbocation, carbon radical or carbanion starts to develop, label where that will occur.

4-methylpenta-1,3-diene with $CH_2=CHCH=O$

b. Use mechanism arrows to illustrate the reaction that occurs.

If applicable, use stabilization resources to deal with the carbocation, carbon radical, or carbanion that starts to develop during the reaction, and draw the structure of any resonance-stabilized intermediate.

Continue labelling and diagramming the reaction until you find the major stable product(s).

Finally, state the stereochemistry of the major product(s) and use either Fisher projection or perspective formula representations to illustrate that stereochemistry.

14. a. Label the reactive features, highlight the most reactive feature, and then highlight what it needs. Also, state if a carbocation, carbon radical, or carbanion will start to develop, and/or if aromatic character will be lost because of a reaction between these molecules. If a carbocation, carbon radical or carbanion starts to develop, label where that will occur.

$(3S,4R)$-4-methylhexan-3-ol with concentrated H_3PO_4

b. Use mechanism arrows to illustrate the reaction that occurs.

If applicable, use stabilization resources to deal with the carbocation, carbon radical, or carbanion that starts to develop during the reaction, and draw the structure of any resonance-stabilized intermediate.

Continue labelling and diagramming the reaction until you find the major stable product(s).

Finally, state the stereochemistry of the major product(s) and use either Fisher projection or perspective formula representations to illustrate that stereochemistry.

15. a. Label the reactive features, highlight the most reactive feature, and then highlight what it needs. Also, state if a carbocation, carbon radical, or carbanion will start to develop, and/or if aromatic character will be lost because of a reaction between these molecules. If a carbocation, carbon radical or carbanion starts to develop, label where that will occur.

$$HN-\overset{\overset{\text{O}}{\|}}{C}-CH_3$$

Br $+\ H_2SO_4$

b. Use mechanism arrows to illustrate the reaction that occurs.

If applicable, use stabilization resources to deal with the carbocation, carbon radical, or carbanion that starts to develop during the reaction, and draw the structure of any resonance-stabilized intermediate.

Continue labelling and diagramming the reaction until you find the major stable product(s).

Finally, state the stereochemistry of the major product(s) and use either Fisher projection or perspective formula representations to illustrate that stereochemistry.

16. a. Label the reactive features, highlight the most reactive feature, and then highlight what it needs. Also, state if a carbocation, carbon radical, or carbanion will start to develop, and/or if aromatic character will be lost because of a reaction between these molecules. If a carbocation, carbon radical or carbanion starts to develop, label where that will occur.

3,3-dimethylhept-1-ene + Br_2 in methanol

b. Use mechanism arrows to illustrate the reaction that occurs.

If applicable, use stabilization resources to deal with the carbocation, carbon radical, or carbanion that starts to develop during the reaction, and draw the structure of any resonance-stabilized intermediate.

Continue labelling and diagramming the reaction until you find the major stable product(s).

Finally, state the stereochemistry of the major product(s) and use either Fisher projection or perspective formula representations to illustrate that stereochemistry.

17. a. Label the reactive features, highlight the most reactive feature, and then highlight what it needs. Also, state if a carbocation, carbon radical, or carbanion will start to develop, and/or if aromatic character will be lost because of a reaction between these molecules. If a carbocation, carbon radical or carbanion starts to develop, label where that will occur.

chlorobenzene + 2-methylpropanoyl chloride + AlCl$_3$

b. Use mechanism arrows to illustrate the reaction that occurs.

If applicable, use stabilization resources to deal with the carbocation, carbon radical, or carbanion that starts to develop during the reaction, and draw the structure of any resonance-stabilized intermediate.

Continue labelling and diagramming the reaction until you find the major stable product(s).

Finally, state the stereochemistry of the major product(s) and use either Fisher projection or perspective formula representations to illustrate that stereochemistry.

18. (Chapters 1, 13, 15, and 16) Analysis of a compound with the molecular formula of $C_6H_{12}O_2$ produced the following IR and NMR data:

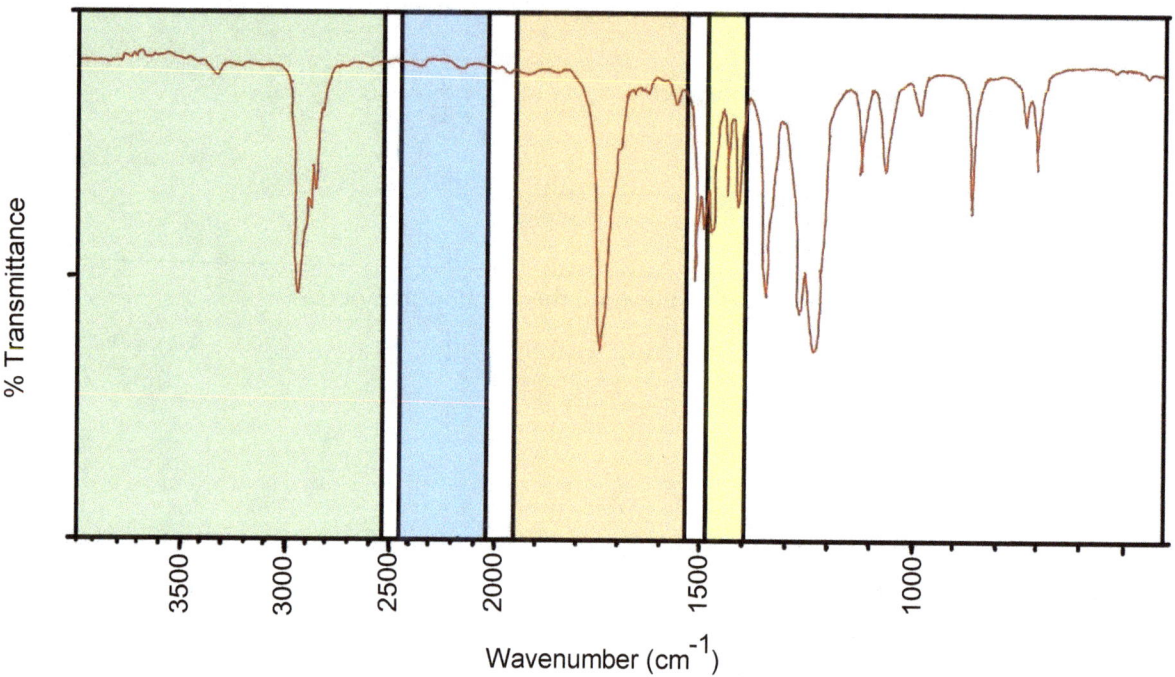

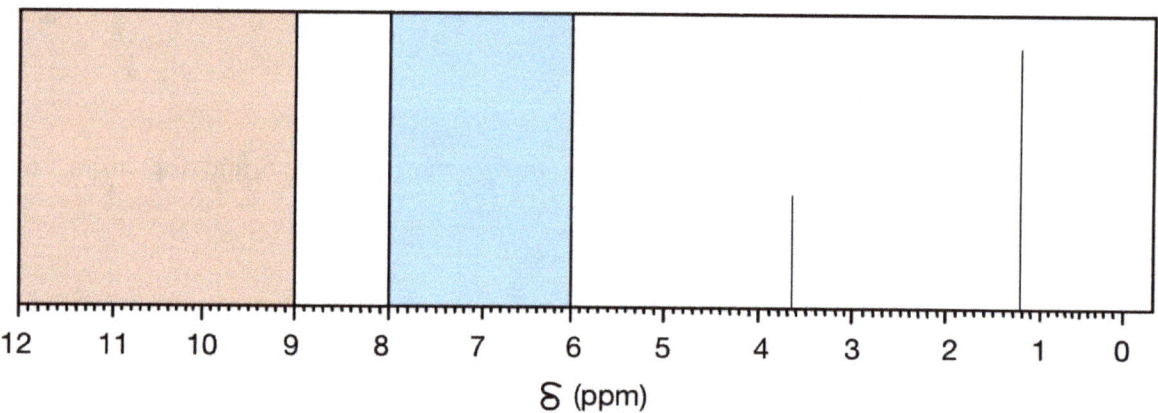

a. Draw the structure of the compound.

b. Write the IUPAC and common name for the compound.

IUPAC: _____

Common name: _____

c. Predict the product of a reaction between this compound and potassium ethoxide.

_____

d. Propose a method for synthesizing this compound starting with any alkyl halide.

_____

19. (Chapters 1, 3, 6, 14, 15, and 16) MS, IR, and NMR analysis of a compound produced the following data:

   MS: M+ 100, base peak 43, other significant peak: 58 m/z
   IR: bands at approximately: 2900, 1730 (deep) cm-1
   NMR: signals at approximately 0.93 (d), 2.12 (s), 2.16 (m), 2.31 (d) ppm

a. Draw the structure of the compound.

   _____

b. Write the IUPAC name for the compound.

   _____

c. Predict the product of an aldol reaction of this compound in the presence of sodium hydroxide.

   _____

Mash-up 4

1. (Chapter 1) Write the IUPAC name for the following compound:

$$H_2C-C\equiv CH$$

$$H_3C-CH_2-CH_2-CH_2-CH-CH_2-CH_2-CH_2-CH_3$$

_____

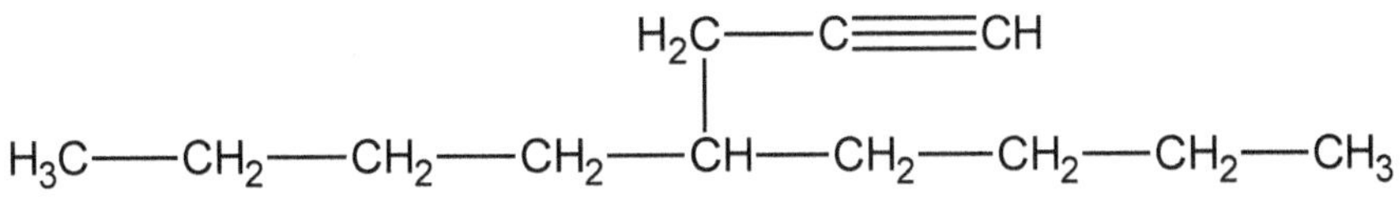

2. (Chapters 1 and 3) Write the IUPAC and the common name for the following compound:

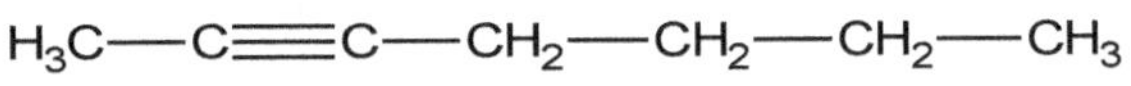

$$H_3C-C\equiv C-CH_2-CH_2-CH_2-CH_3$$

IUPAC: _____

$$H_3C-C\equiv C-CH_2-CH_2-CH_2-CH_3$$

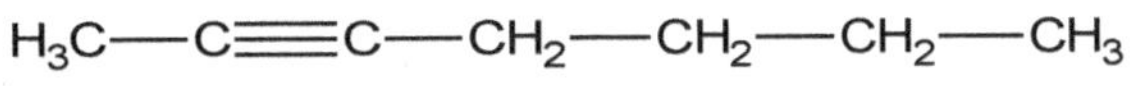

Common name: _____

3. (Chapter 11) Write the IUPAC and the common name for the following compound:

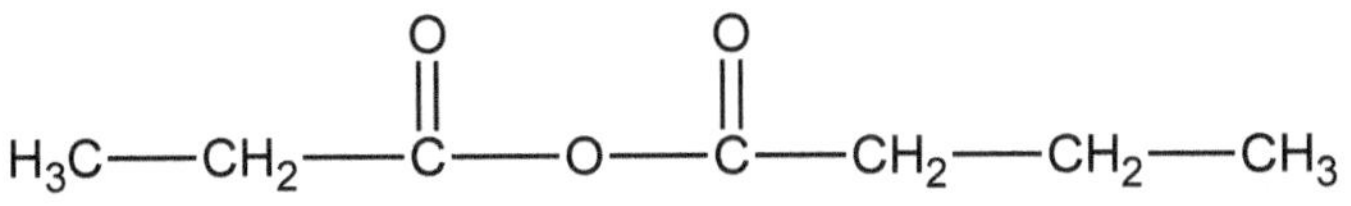

$$H_3C-CH_2-\overset{\overset{\displaystyle O}{\|}}{C}-O-\overset{\overset{\displaystyle O}{\|}}{C}-CH_2-CH_2-CH_3$$

IUPAC: _____

$$H_3C-CH_2-\overset{\overset{\displaystyle O}{\|}}{C}-O-\overset{\overset{\displaystyle O}{\|}}{C}-CH_2-CH_2-CH_3$$

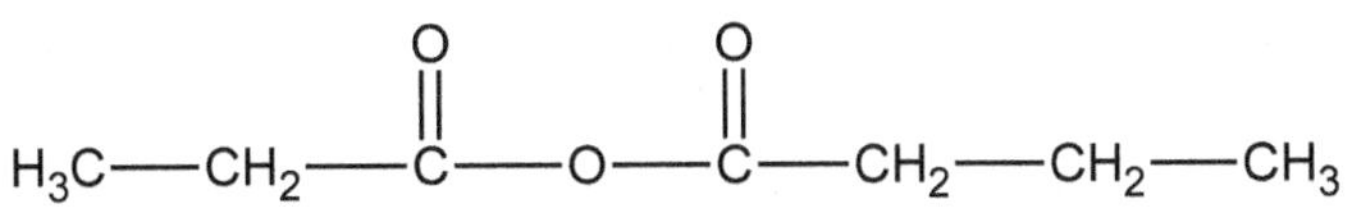

Common name: _____

4. (Chapter 10) Write the common name for the following compound:

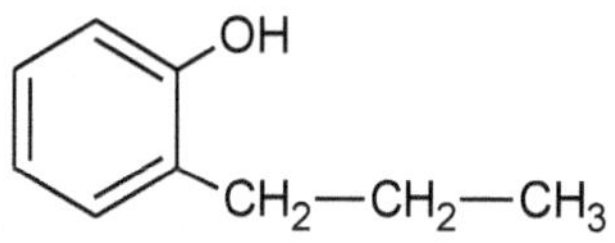

_____

5. (Chapter 15) If you saw a band on an IR spectrum at approximately 1750, where else would you look to determine what type of carbonyl compound the molecule has?

_____

_____

6. (Chapter 12) What type of instability will need to be addressed in the intermediate that is created when a carbonyl carbon reacts with a nucleophile/base?

_____

7. (Chapter 9) What general resource could potentially be used to stabilize a carbocation when aromatic character needs to be restored?

_____

_____

_____

_____

_____

(Chapters 5–13) For each of the following:

8. a. Label the reactive features, highlight the most reactive feature, and then highlight what it needs. Also, state if a carbocation, carbon radical, or carbanion will start to develop, and/or if aromatic character will be lost because of a reaction between these molecules. If a carbocation, carbon radical or carbanion starts to develop, label where that will occur.

    3-bromo-3-methylpentane + NaOH

    b. Use mechanism arrows to illustrate the reaction that occurs.

    If applicable, use stabilization resources to deal with the carbocation, carbon radical, or carbanion that starts to develop during the reaction, and draw the structure of any resonance-stabilized intermediate.

    Continue labelling and diagramming the reaction until you find the major stable product(s).

    Finally, state the stereochemistry of the major product(s) and use either Fisher projection or perspective formula representations to illustrate that stereochemistry.

9. a. Label the reactive features, highlight the most reactive feature, and then highlight what it needs. Also, state if a carbocation, carbon radical, or carbanion will start to develop, and/or if aromatic character will be lost because of a reaction between these molecules. If a carbocation, carbon radical or carbanion starts to develop, label where that will occur.

acetophenone + H_2SO_4

b. Use mechanism arrows to illustrate the reaction that occurs.

If applicable, use stabilization resources to deal with the carbocation, carbon radical, or carbanion that starts to develop during the reaction, and draw the structure of any resonance-stabilized intermediate.

Continue labelling and diagramming the reaction until you find the major stable product(s).

Finally, state the stereochemistry of the major product(s) and use either Fisher projection or perspective formula representations to illustrate that stereochemistry.

10. a. Label the reactive features, highlight the most reactive feature, and then highlight what it needs. Also, state if a carbocation, carbon radical, or carbanion will start to develop, and/or if aromatic character will be lost because of a reaction between these molecules. If a carbocation, carbon radical or carbanion starts to develop, label where that will occur.

pent-1-yne + 2 equivalents of Cl_2

b. Use mechanism arrows to illustrate the reaction that occurs.

If applicable, use stabilization resources to deal with the carbocation, carbon radical, or carbanion that starts to develop during the reaction, and draw the structure of any resonance-stabilized intermediate.

Continue labelling and diagramming the reaction until you find the major stable product(s).

Finally, state the stereochemistry of the major product(s) and use either Fisher projection or perspective formula representations to illustrate that stereochemistry.

11. a. Label the reactive features, highlight the most reactive feature, and then highlight what it needs. Also, state if a carbocation, carbon radical, or carbanion will start to develop, and/or if aromatic character will be lost because of a reaction between these molecules. If a carbocation, carbon radical or carbanion starts to develop, label where that will occur.

isobutyl dimethyl propyl amine + NaOH

b. Use mechanism arrows to illustrate the reaction that occurs.

If applicable, use stabilization resources to deal with the carbocation, carbon radical, or carbanion that starts to develop during the reaction, and draw the structure of any resonance-stabilized intermediate.

Continue labelling and diagramming the reaction until you find the major stable product(s).

Finally, state the stereochemistry of the major product(s) and use either Fisher projection or perspective formula representations to illustrate that stereochemistry.

12. a. Label the reactive features, highlight the most reactive feature, and then highlight what it needs. Also, state if a carbocation, carbon radical, or carbanion will start to develop, and/or if aromatic character will be lost because of a reaction between these molecules. If a carbocation, carbon radical or carbanion starts to develop, label where that will occur.

m-chloropropylbenzene + H_2SO_4

b. Use mechanism arrows to illustrate the reaction that occurs.

If applicable, use stabilization resources to deal with the carbocation, carbon radical, or carbanion that starts to develop during the reaction, and draw the structure of any resonance-stabilized intermediate.

Continue labelling and diagramming the reaction until you find the major stable product(s).

Finally, state the stereochemistry of the major product(s) and use either Fisher projection or perspective formula representations to illustrate that stereochemistry.

13. a. Label the reactive features, highlight the most reactive feature, and then highlight what it needs. Also, state if a carbocation, carbon radical, or carbanion will start to develop, and/or if aromatic character will be lost because of a reaction between these molecules. If a carbocation, carbon radical or carbanion starts to develop, label where that will occur.

1-methylcyclopenta-1,3-diene with $CH_2=CHC\equiv N$

b. Use mechanism arrows to illustrate the reaction that occurs.

If applicable, use stabilization resources to deal with the carbocation, carbon radical, or carbanion that starts to develop during the reaction, and draw the structure of any resonance-stabilized intermediate.

Continue labelling and diagramming the reaction until you find the major stable product(s).

Finally, state the stereochemistry of the major product(s) and use either Fisher projection or perspective formula representations to illustrate that stereochemistry.

14. a. Label the reactive features, highlight the most reactive feature, and then highlight what it needs. Also, state if a carbocation, carbon radical, or carbanion will start to develop, and/or if aromatic character will be lost because of a reaction between these molecules. If a carbocation, carbon radical or carbanion starts to develop, label where that will occur.

phenylacetate + 2-chloro-3-methyl butane with $AlCl_3$

b. Use mechanism arrows to illustrate the reaction that occurs.

If applicable, use stabilization resources to deal with the carbocation, carbon radical, or carbanion that starts to develop during the reaction, and draw the structure of any resonance-stabilized intermediate.

Continue labelling and diagramming the reaction until you find the major stable product(s).

Finally, state the stereochemistry of the major product(s) and use either Fisher projection or perspective formula representations to illustrate that stereochemistry.

15. a. Label the reactive features, highlight the most reactive feature, and then highlight what it needs. Also, state if a carbocation, carbon radical, or carbanion will start to develop, and/or if aromatic character will be lost because of a reaction between these molecules. If a carbocation, carbon radical or carbanion starts to develop, label where that will occur.

       (Z)-3-methylhept-3-ene + Cl_2 in CH_2Cl_2

b. Use mechanism arrows to illustrate the reaction that occurs.

If applicable, use stabilization resources to deal with the carbocation, carbon radical, or carbanion that starts to develop during the reaction, and draw the structure of any resonance-stabilized intermediate.

Continue labelling and diagramming the reaction until you find the major stable product(s).

Finally, state the stereochemistry of the major product(s) and use either Fisher projection or perspective formula representations to illustrate that stereochemistry.

16. (Chapters 1, 13, 15, and 16) A compound with the molecular formula of $C_9H_{10}O$ produced the following IR and NMR data:

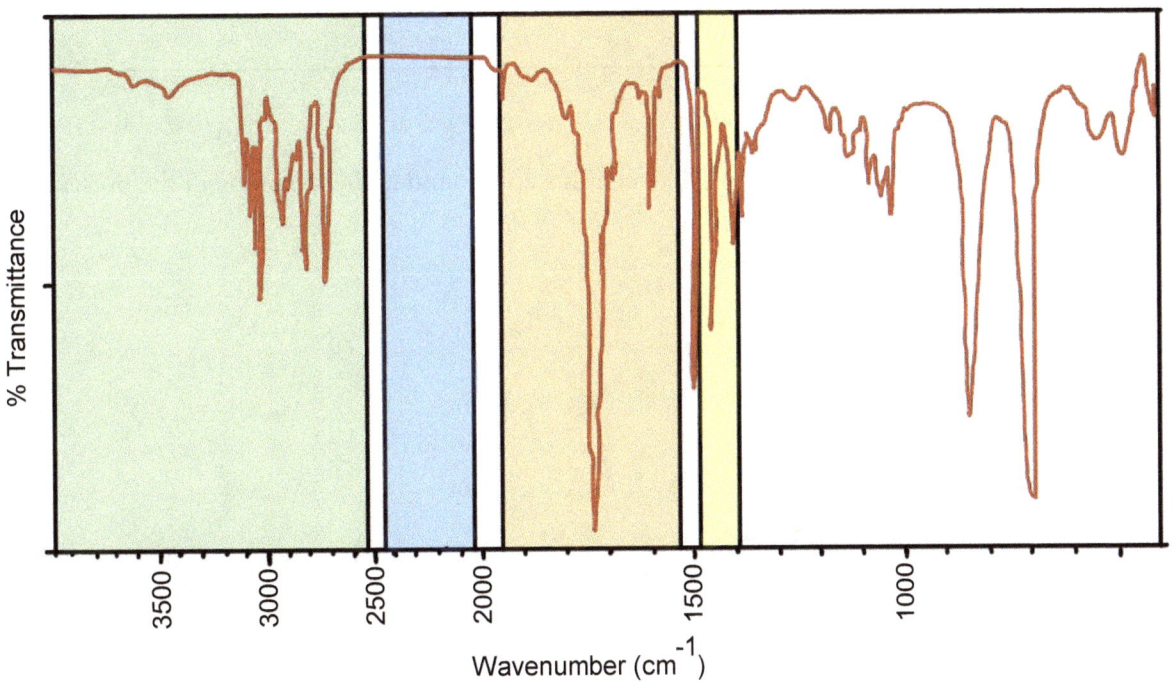

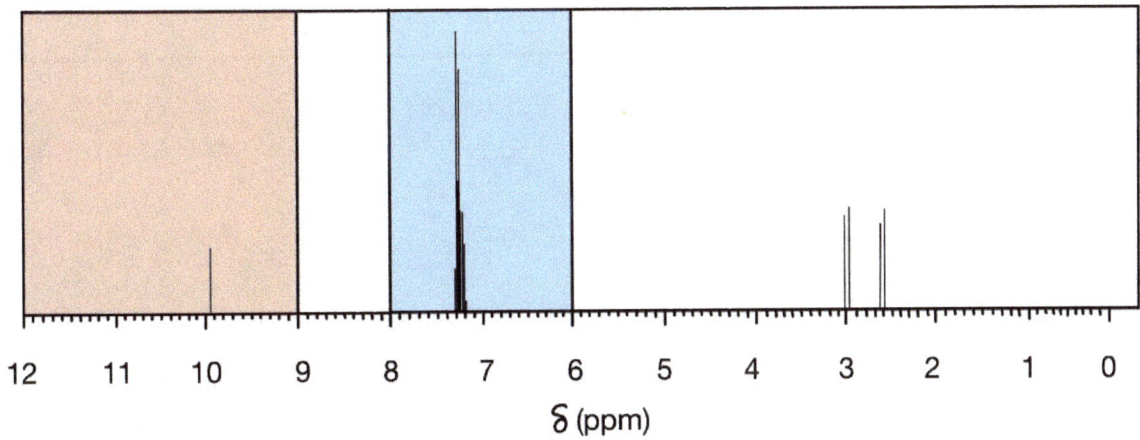

a. Draw the structure of the compound.

b. Write the IUPAC name for the compound.

_____

c. Predict the product of a reaction between this compound and ethylamine in the presence of a trace of acid.

_____

17. (Chapters 1, 3, 6, 14, 15, and 16) MS, IR, and NMR analysis of a compound produced the following data:

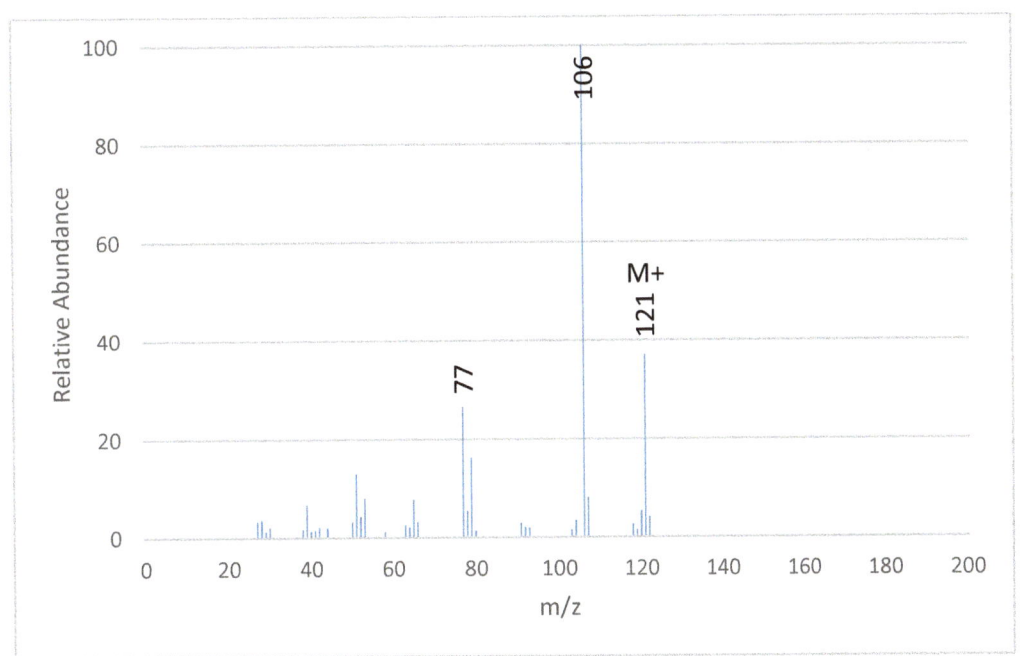

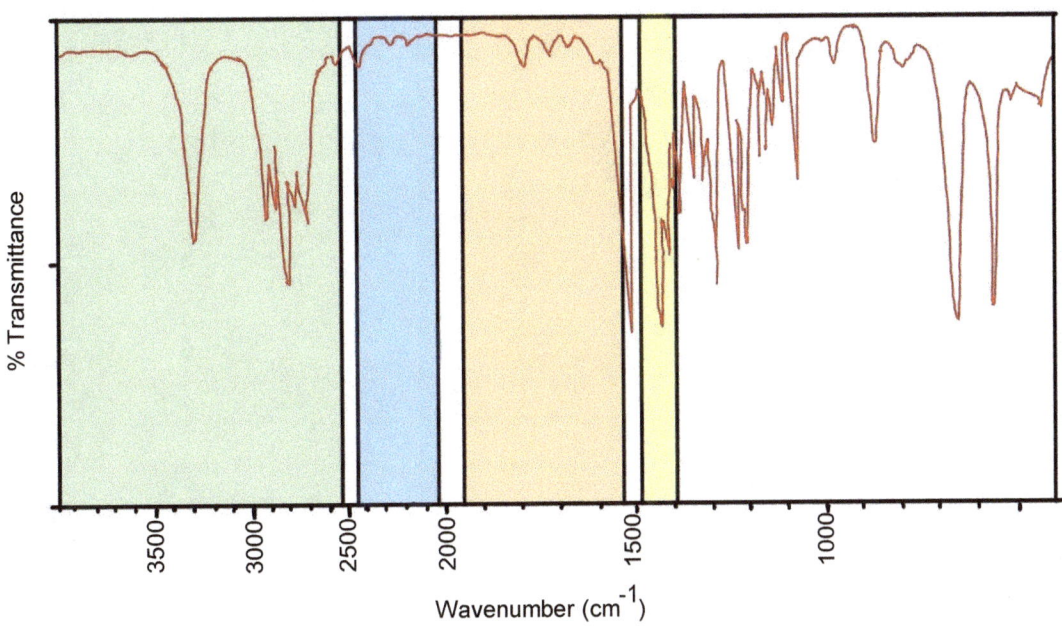

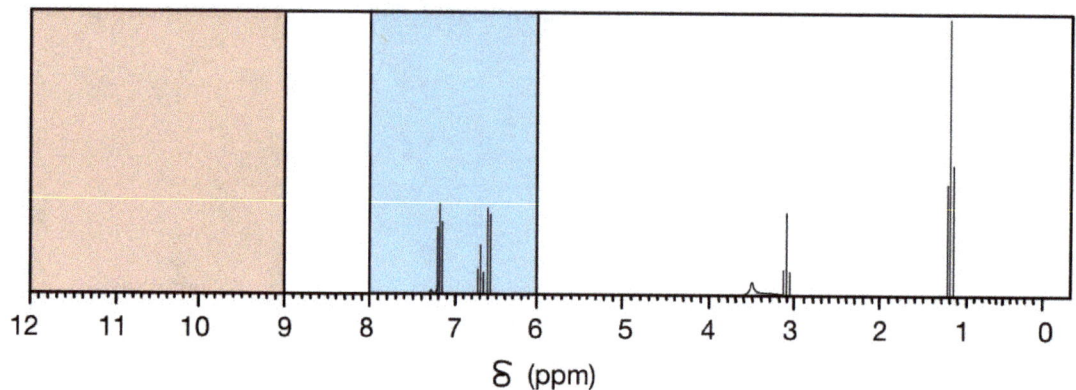

a. Draw the structure of the compound.

_____

b. Write the IUPAC name for the compound.

_____

Mash-up 5

1. (Chapters 1 and 3) Write the IUPAC and the common name for the following compound:

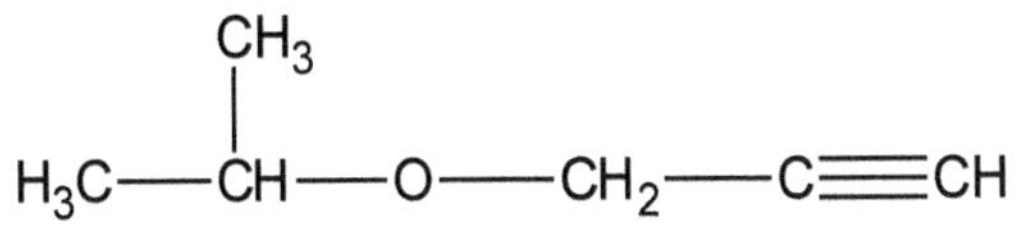

$$H_3C-\underset{\underset{\displaystyle CH_3}{|}}{CH}-O-CH_2-C\equiv CH$$

IUPAC: _____

$$H_3C-\underset{\underset{\displaystyle CH_3}{|}}{CH}-O-CH_2-C\equiv CH$$

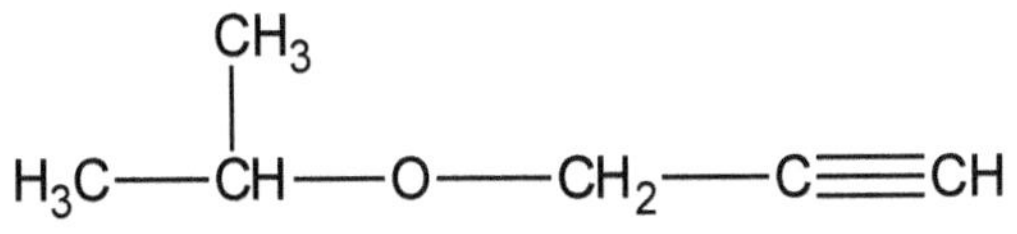

Common name: _____

2. (Chapters 1 and 3) Write the IUPAC and the common name for the following compound:

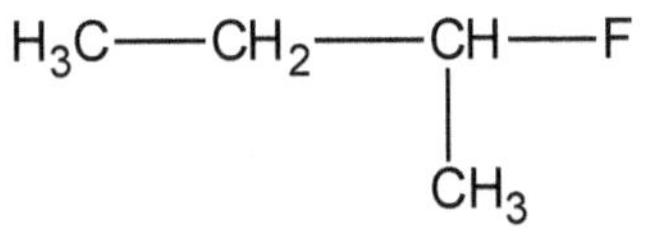

$$H_3C-CH_2-\underset{\underset{\displaystyle CH_3}{|}}{CH}-F$$

IUPAC: _____

$$H_3C-CH_2-\underset{\underset{\displaystyle CH_3}{|}}{CH}-F$$

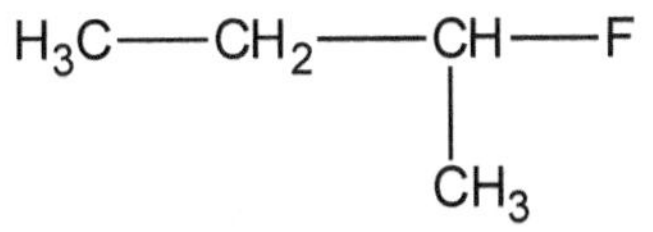

Common name: _____

3. (Chapter 10) Write the common name for the following compound:

Br — C₆H₄ — $CH=CH_2$

_____

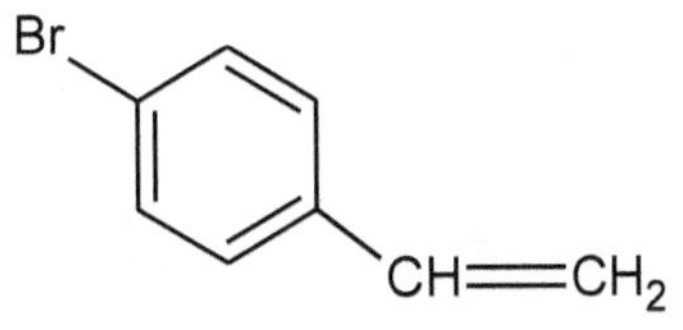

4. (Chapter 11) Write the common name for the following compound:

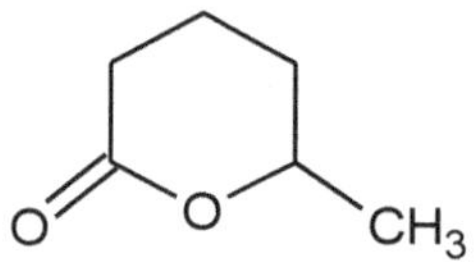

_____

5. (Chapter 15) Where would you look on an IR chart to double check and confirm the identity of a band at 3300?

_____

6. (Chapter 9) What type of instability will need to be addressed in the intermediate that is created when an enol reacts under acidic conditions?

_____

7. (Chapters 6, 7, and 8) What general methods could potentially be used to stabilize a carbocation?

_____

_____

_____

_____

_____

8. (Chapter 15) If you needed to determine the identity of a proton NMR signal in the 6–8 ppm range, where would you look on an IR chart to distinguish whether it came from a vinyl hydrogen or an aryl hydrogen?

_____

_____

(Chapters 5–13) For each of the following:

9.   a. Label the reactive features, highlight the most reactive feature, and then highlight what it needs. Also, state if a carbocation, carbon radical, or carbanion will start to develop, and/or if aromatic character will be lost because of a reaction between these molecules. If a carbocation, carbon radical or carbanion starts to develop, label where that will occur.

(E)-4,5-dimethyloct-4-ene + BH_3 followed by H_2O_2, ^-OH, and H_2O

b. Use mechanism arrows to illustrate the reaction that occurs.

If applicable, use stabilization resources to deal with the carbocation, carbon radical, or carbanion that starts to develop during the reaction, and draw the structure of any resonance-stabilized intermediate.

Continue labelling and diagramming the reaction until you find the major stable product(s).

Finally, state the stereochemistry of the major product(s) and use either Fisher projection or perspective formula representations to illustrate that stereochemistry.

10. a. Label the reactive features, highlight the most reactive feature, and then highlight what it needs. Also, state if a carbocation, carbon radical, or carbanion will start to develop, and/or if aromatic character will be lost because of a reaction between these molecules. If a carbocation, carbon radical or carbanion starts to develop, label where that will occur.

   (3S,4R)-4-methylhexan-3-ol + TsCl and NaSH

   b. Use mechanism arrows to illustrate the reaction that occurs.

   If applicable, use stabilization resources to deal with the carbocation, carbon radical, or carbanion that starts to develop during the reaction, and draw the structure of any resonance-stabilized intermediate.

   Continue labelling and diagramming the reaction until you find the major stable product(s).

   Finally, state the stereochemistry of the major product(s) and use either Fisher projection or perspective formula representations to illustrate that stereochemistry.

11. a. Label the reactive features, highlight the most reactive feature, and then highlight what it needs. Also, state if a carbocation, carbon radical, or carbanion will start to develop, and/or if aromatic character will be lost because of a reaction between these molecules. If a carbocation, carbon radical or carbanion starts to develop, label where that will occur.

toluene + HNO₃ with tr. H₂SO₄

b. Use mechanism arrows to illustrate the reaction that occurs.

If applicable, use stabilization resources to deal with the carbocation, carbon radical, or carbanion that starts to develop during the reaction, and draw the structure of any resonance-stabilized intermediate.

Continue labelling and diagramming the reaction until you find the major stable product(s).

Finally, state the stereochemistry of the major product(s) and use either Fisher projection or perspective formula representations to illustrate that stereochemistry.

12. a. Label the reactive features, highlight the most reactive feature, and then highlight what it needs. Also, state if a carbocation, carbon radical, or carbanion will start to develop, and/or if aromatic character will be lost because of a reaction between these molecules. If a carbocation, carbon radical or carbanion starts to develop, label where that will occur.

2,3-dimethylhepta-1,3-diene with Br_2

b. Use mechanism arrows to illustrate the reaction that occurs.

If applicable, use stabilization resources to deal with the carbocation, carbon radical, or carbanion that starts to develop during the reaction, and draw the structure of any resonance-stabilized intermediate.

Continue labelling and diagramming the reaction until you find the major stable product(s).

Finally, state the stereochemistry of the major product(s) and use either Fisher projection or perspective formula representations to illustrate that stereochemistry.

13. a. Label the reactive features, highlight the most reactive feature, and then highlight what it needs. Also, state if a carbocation, carbon radical, or carbanion will start to develop, and/or if aromatic character will be lost because of a reaction between these molecules. If a carbocation, carbon radical or carbanion starts to develop, label where that will occur.

p-isopropylphenol + 2-chloro-3-methyl butane + AlCl₃

b. Use mechanism arrows to illustrate the reaction that occurs.

If applicable, use stabilization resources to deal with the carbocation, carbon radical, or carbanion that starts to develop during the reaction, and draw the structure of any resonance-stabilized intermediate.

Continue labelling and diagramming the reaction until you find the major stable product(s).

Finally, state the stereochemistry of the major product(s) and use either Fisher projection or perspective formula representations to illustrate that stereochemistry.

14. a. Label the reactive features, highlight the most reactive feature, and then highlight what it needs. Also, state if a carbocation, carbon radical, or carbanion will start to develop, and/or if aromatic character will be lost because of a reaction between these molecules. If a carbocation, carbon radical or carbanion starts to develop, label where that will occur.

1-methylcyclopentene + HBr and H_2O_2

b. Use mechanism arrows to illustrate the reaction that occurs.

If applicable, use stabilization resources to deal with the carbocation, carbon radical, or carbanion that starts to develop during the reaction, and draw the structure of any resonance-stabilized intermediate.

Continue labelling and diagramming the reaction until you find the major stable product(s).

Finally, state the stereochemistry of the major product(s) and use either Fisher projection or perspective formula representations to illustrate that stereochemistry.

15. a. Label the reactive features, highlight the most reactive feature, and then highlight what it needs. Also, state if a carbocation, carbon radical, or carbanion will start to develop, and/or if aromatic character will be lost because of a reaction between these molecules. If a carbocation, carbon radical or carbanion starts to develop, label where that will occur.

HN—CH₃

+ Cl₂ and FeCl₃

b. Use mechanism arrows to illustrate the reaction that occurs.

If applicable, use stabilization resources to deal with the carbocation, carbon radical, or carbanion that starts to develop during the reaction, and draw the structure of any resonance-stabilized intermediate.

Continue labelling and diagramming the reaction until you find the major stable product(s).

Finally, state the stereochemistry of the major product(s) and use either Fisher projection or perspective formula representations to illustrate that stereochemistry.

16. a. Label the reactive features, highlight the most reactive feature, and then highlight what it needs. Also, state if a carbocation, carbon radical, or carbanion will start to develop, and/or if aromatic character will be lost because of a reaction between these molecules. If a carbocation, carbon radical or carbanion starts to develop, label where that will occur.

   isobutyl methyl propyl amine with H_2O_2

   b. Use mechanism arrows to illustrate the reaction that occurs.

   If applicable, use stabilization resources to deal with the carbocation, carbon radical, or carbanion that starts to develop during the reaction, and draw the structure of any resonance-stabilized intermediate.

   Continue labelling and diagramming the reaction until you find the major stable product(s).

   Finally, state the stereochemistry of the major product(s) and use either Fisher projection or perspective formula representations to illustrate that stereochemistry.

17. (Chapters 1, 13, 15, and 16) A compound with the molecular formula of $C_4H_8O_2$ produces the following IR and NMR spectra:

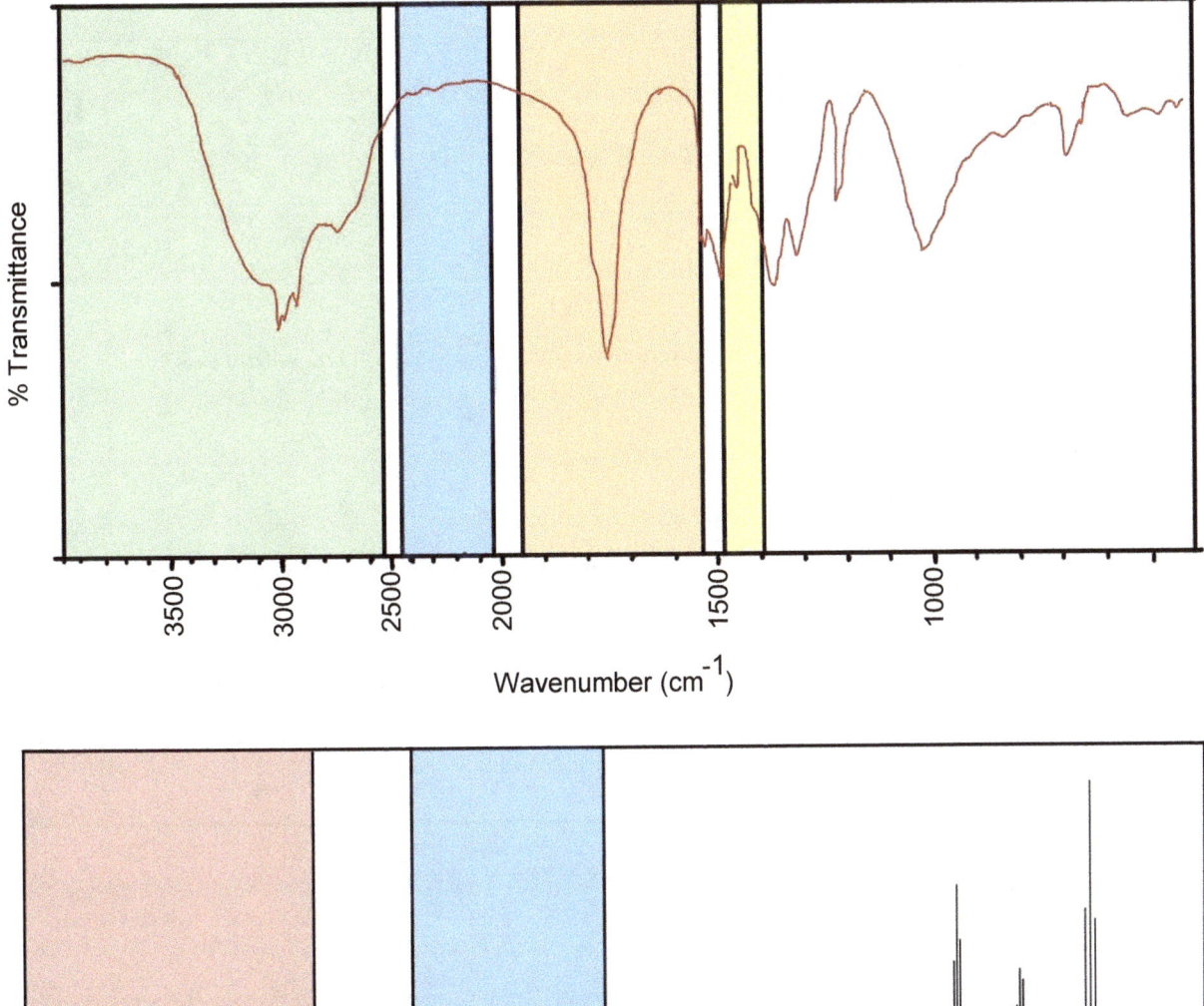

a. Draw the structure of the compound.

b. Write the IUPAC and common names for the compound.

IUPAC: _____

Common Name: _____

c. Predict the product of the reaction when the compound is reacted with TsCl.

_____

18. (Chapters 1, 3, 6, 14, 15, and 16) MS, IR, and NMR analysis of a compound produced the following data:

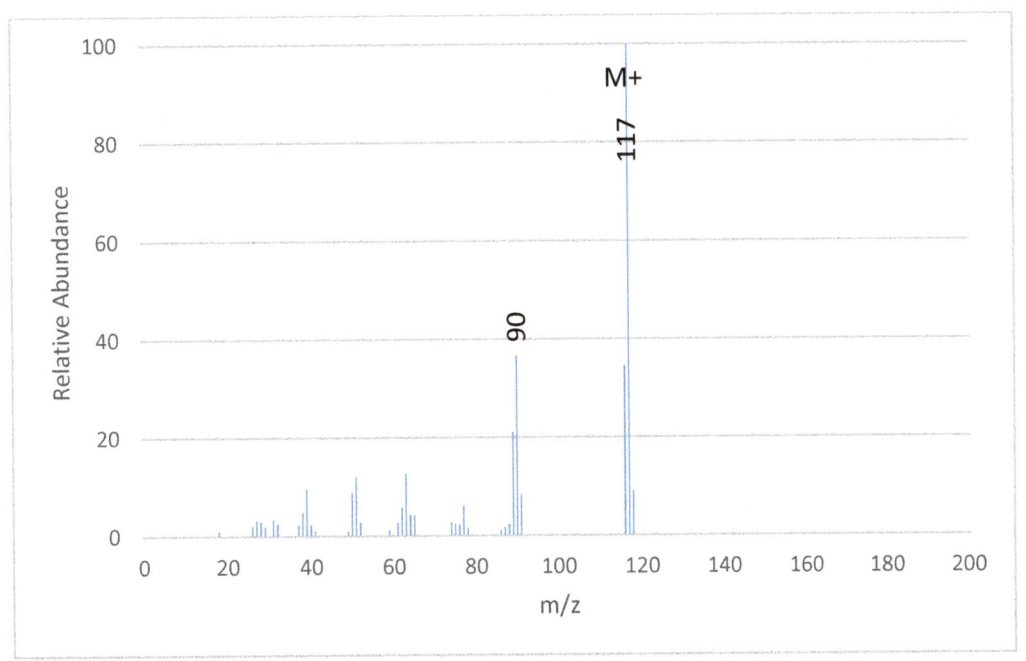

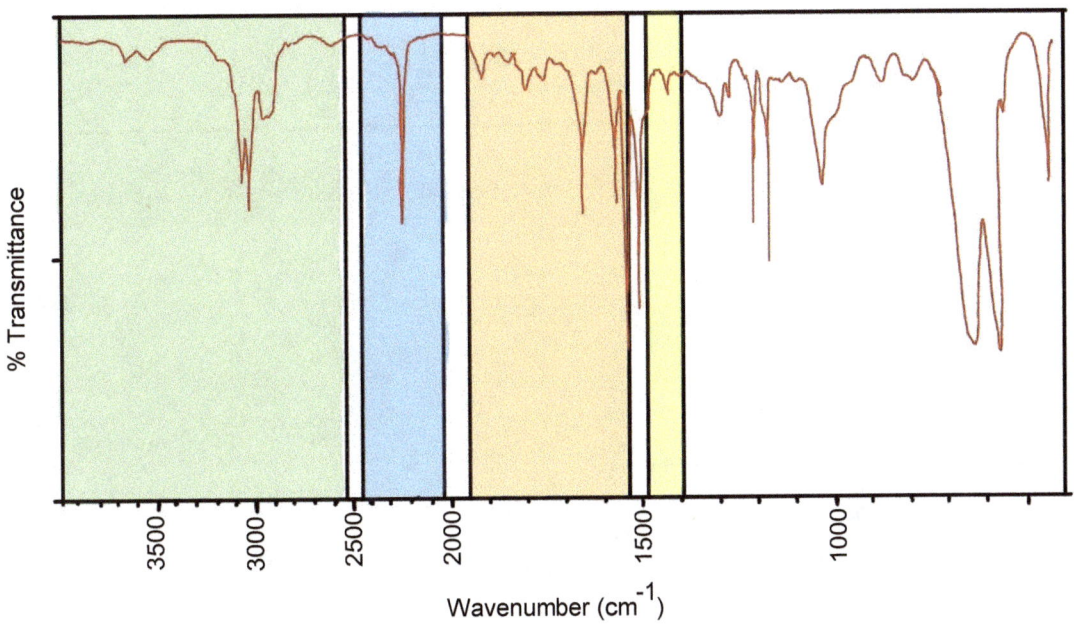

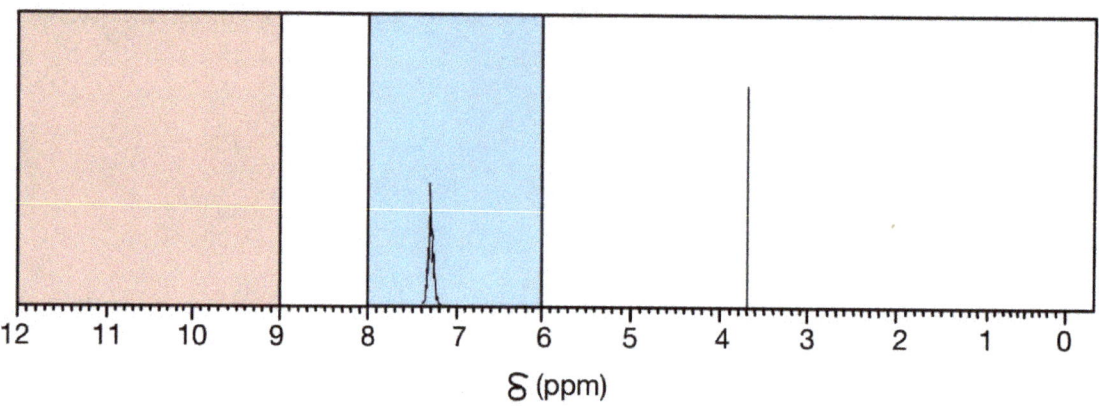

a. Draw the structure of the compound.

_____

b. Write the common name for the compound.

_____

Chapter 16

Summary of Concepts and Analysis Methods

Summary of the Chapter:

_____

_____

_____

_____

_____

_____

_____

_____

Questions You Should Ask In Class:

_____

_____

_____

_____

_____

Common Mistakes You Tend to Make but Want to Avoid in the Future:

_____

_____

_____

_____

_____

Types of Problems Needed for Targeted Practice:

_____

_____

_____

_____

_____

_____

Chapter 17

Understanding Organic Redox Reactions and Preparing for the ACS Exam

Key Concepts

When the net number of bonds to OXYGEN or a halogen INCREASE, or when the net number of bonds to hydrogen decrease, a molecule is OXIDIZED. When the net number of bonds to oxygen or a halogen is REDUCED (or the number of bonds to hydrogen increase), the substance is REDUCED.

A recognizable pattern for many common oxidizing agents is that they have a significant number of oxygen atoms. Therefore, the central atom has a very positive oxidation number. (Examples include H_2CrO_4, $NaCr_2O_7$, $KMnO_4$, OsO_4, and HIO_4.) When a large central atom (one with 5 or 6 shells) has a high oxidation number, it oxidizes an organic molecule to the next highest oxidation state. When the oxidizing agent contains a smaller central atom with a high oxidation number, it is more aggressive, so it moves a sample 2 oxidation levels.

What You Need To Learn, Understand, and Apply

1. How to determine when an organic reaction involves oxidation or reduction. (pages 504–506)
2. How to predict the outcomes of reduction reactions commonly used in organic chemistry, including those that use each of the following reactants: H_2 with Pt, Pd or Ni; Lindlar's reagent; Raney nickel; Na or Li with NH_3; $NaBH_4$ or $LiAlH_4$; and NH_2NH_2 with hydroxide. (pages 507–513)
3. How to predict the outcomes of oxidation reactions commonly used in organic chemistry, including those that use each of the following reactants: Chromium(VI) (such as $Na_2Cr_2O_7$, H_2CrO_4, and PCC); Manganese(VII) ($KMnO_4$); I(VII) (HIO_4); Osmium(VIII) (OsO_4); and O_3. (pages 514–517)
4. How to protect a functional group from oxidation while reacting another. (pages 518–519)
5. How to predict radical-based reactions of alkanes/alkyl groups. (pages 519–520)
6. How to work multi-step synthesis reactions. (pages 521–524)
7. How to prepare for the ACS Examination in Organic Chemistry. (pages 524–527)

Chapter Preview

The general purpose of this chapter:

_____

_____

_____

_____

_____

(Keep this purpose in mind as you read the chapter to help you tie all the concepts together into one complete picture.)

Questions to Assess Understanding	Lecture/Reading
	After filling in your lecture note outline, fold the paper over so that only the assessment question is visible. Once you can consistently answer a given question correctly on your own, place an X by that question.
	Learning Objective 1: Be able to determine when an organic reaction involves oxidation or reduction. (pages 504–506)
____ How can you easily tell that a reaction oxidized an organic molecule?	How to easily tell that a reaction oxidized an organic molecule:
____ How can you easily tell that a reaction reduced an organic molecule?	How to easily tell that a reaction reduced an organic molecule:
____ What are the stages of oxidation?	Oxidation Stages:

Skills Check 1

Has the organic molecule been oxidized or reduced in each of the following reactions?

a.   Alcohol to ketone

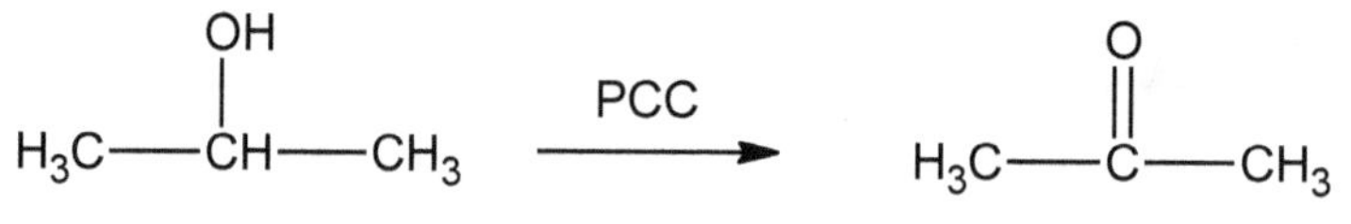

$$H_3C-\underset{\underset{OH}{|}}{CH}-CH_3 \xrightarrow{PCC} H_3C-\underset{\underset{\parallel}{O}}{C}-CH_3$$

———————————————

b.   Alcohol to carboxylic acid

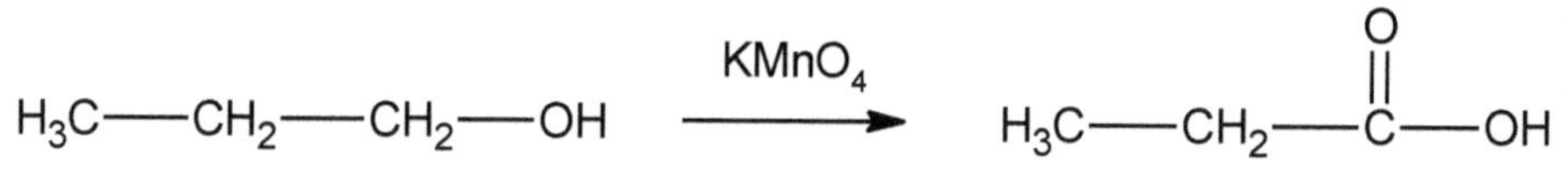

$$H_3C-CH_2-CH_2-OH \xrightarrow{KMnO_4} H_3C-CH_2-\underset{\overset{\parallel}{O}}{C}-OH$$

———————————————

c.   Ketone to alcohol

$$H_3C-CH_2-\underset{\overset{\parallel}{O}}{C}-H \xrightarrow[\text{2. acid}]{\text{1. NaBH}_4} H_3C-CH_2-CH_2-OH$$

———————————————

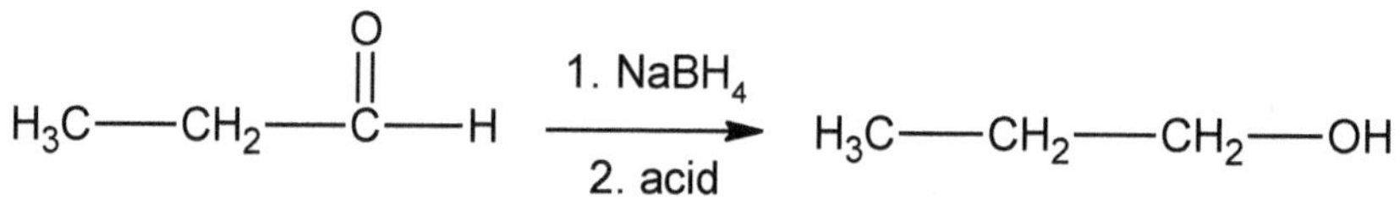

d. Alkene to diol

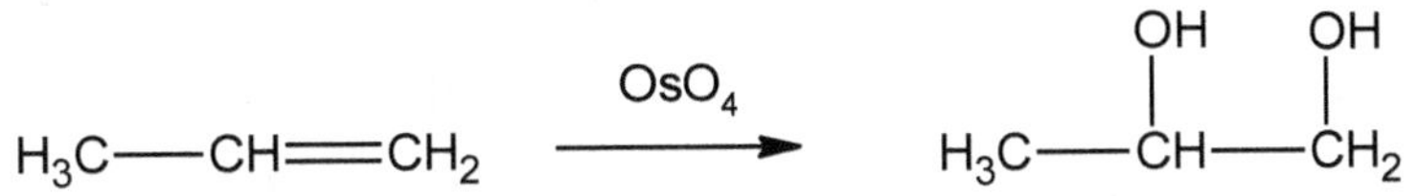

$$H_3C-CH=CH_2 \xrightarrow{OsO_4} H_3C-\overset{\overset{\displaystyle OH}{|}}{CH}-\overset{\overset{\displaystyle OH}{|}}{CH_2}$$

_____

e. Alkane to alkyl halide

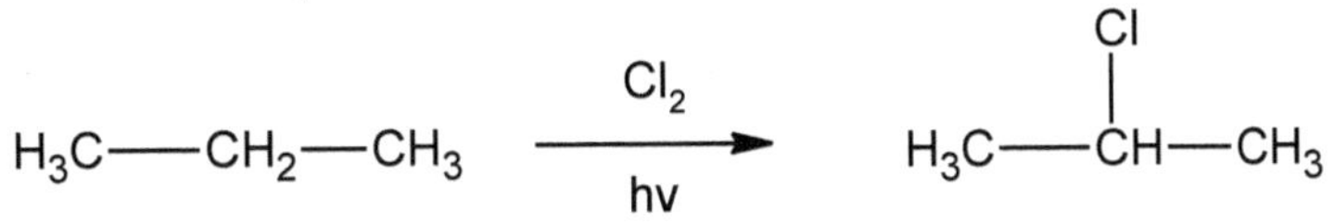

$$H_3C-CH_2-CH_3 \xrightarrow[hv]{Cl_2} H_3C-\overset{\overset{\displaystyle Cl}{|}}{CH}-CH_3$$

_____

f. Ketone to alkane

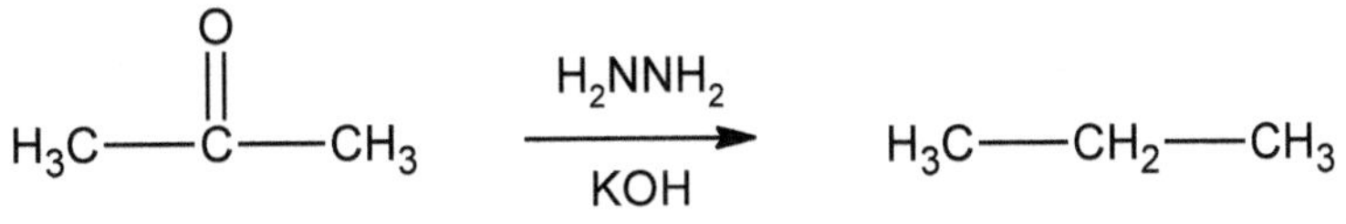

$$H_3C-\overset{\overset{\displaystyle O}{||}}{C}-CH_3 \xrightarrow[KOH]{H_2NNH_2} H_3C-CH_2-CH_3$$

_____

Learning Objective 2: Be able to predict the outcomes of reduction reactions commonly used in organic chemistry, including those that use each of the following reactants: H_2 with Pt, Pd, or Ni; Lindlar's reagent; Raney nickel; Na or Li with NH_3; $NaBH_4$ or $LiAlH_4$; and NH_2NH_2 with hydroxide. (pages 507–513)

Reducing Agents That Create Radicals

Hydrogen Radicals Created by a Catalyst

_____ What is the definition of a hydrogenation reaction?

The definition of a hydrogenation reaction: _____

_____

_____ What is the general mechanism for reactions of a pi bond with H_2 that use Pt, Pd, Ni, Lindlar's reagent, or Raney nickel as the catalyst?

The general mechanism for reactions of a pi bond with H_2 that use Pt, Pd, Ni, Lindlar's reagent, or Raney nickel as the catalyst:

_____ What is Lindlar's catalyst, and what is it used for?

What Lindlar's catalyst is and what it is specifically used for: _____

_____

_____

_____ What is Raney Ni, and what is it used for?

What Raney Ni is and what it is specifically used for: _____

_____

_____

Radicals Created by Elemental Sodium or Lithium

Skills Check 2

_____ What is the mechanism for a reaction between an alkyne and elemental metal, and why does each step occur?

Predict the reactive intermediate formed as a result of a reaction between the following:

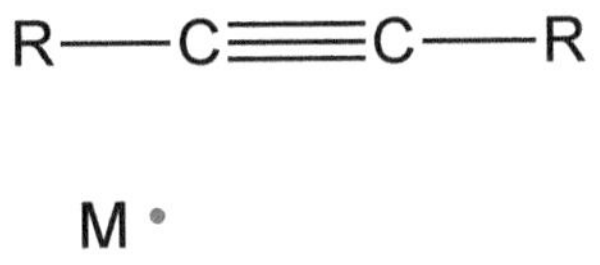

R—C≡C—R

M •

(where M˙ stands for elemental sodium or lithium)

Skills Check 3

Use the skills you have learned so far to work out a resolution for the reactive intermediate. Read the explanation only after you have tried to work the problem yourself. (That is the only effective way to learn the reaction. It's also another opportunity for you to practice applying skills to new situations.)

Here is a list of all the potential resources in the solution: (Note: in the past, we have always crossed off the hydrogens attached to a nitrogen since the bond was too strong for a reaction to occur. This is an unusual situation, however, where the carbanion is such a strong base that it is strong enough to rip a hydrogen away from the nitrogen.)

$$R-C=C-R$$

$$M^+ \quad M^{\cdot} \quad H-N-H \quad H-N-H$$

Reducing Agents That Create a Hydride Ion

_____ What reactants create a hydride ion, and what are they used for?

Reactants that create a hydride ion, and their uses: _____

_____

_____

_____ What is the reason why LiAlH₄ is more reactive than NaBH₄, and when must it specifically be used instead of NaBH₄?

The reason why $LiAlH_4$ is more reactive than $NaBH_4$ and when it must specifically be used instead of $NaBH_4$:

_____

_____

_____

Wolff-Kishner Reaction

Skills Check 4

_____ What is the mechanism for a Wolff-Kishner reaction (the reduction of a carbonyl group to an alkyl carbon), and why does each step occur?

Reason your way through a reaction between the carbonyl group of acetone and NH_2NH_2 (Hint: Treat this like any other reaction between a nitrogen-containing nucleophile and a ketone.) Once you know the product of that reaction, predict the mechanism used to turn the intermediate into propane by reacting it with an aqueous solution of hydroxide.

Learning Objective 3: Be able to predict the outcomes of oxidation reactions commonly used in organic chemistry, including those that use each of the following reactants: Chromium(VI) (such as $Na_2Cr_2O_7$, H_2CrO_4, and PCC); Manganese(VII) ($KMnO_4$); Iodine(VII) (HIO_4); , Osmium(VIII) (OsO_4); and O_3. (pages 514–517)

____ What is the characteristic of most oxidizing agents commonly used in organic reactions?

The characteristic of most oxidizing agents commonly used in organic reactions:

_____

As a refresher, the rules for calculating the oxidation number of an atom in an ionic compound:

1. _____

_____

2. _____

____ How are oxidation numbers calculated?

_____

3. _____

_____

Skills Check 5

1. Calculate the oxidation number of the central atom in each of the following commonly-used oxidizing agents:

   a. Chromium in $Na_2Cr_2O_7$ _____

   b. Chromium in H_2CrO_4

   _____

c. Manganese in $KMnO_4$

_____

d. Iodine in HIO_4

_____

e. Osmium in OsO_4

_____

2. Which is the stronger oxidizing agent within each set?

a. $KMnO_4$ or MnO_2 b. OsO_4 or Ag_2O

_____ What are three specific types of bonds that oxidizing agents generally react with?

Oxidizing agents generally react with fairly weak bonds, including:

1. _____

2. _____

3. _____

The pattern to help you remember which oxidizing agents do what:

1. _____

_____

2. _____

_____

3. _____

_____

___ Why is PCC a mild oxidizing agent?	The reason why PCC is a mild oxidizing agent: _____
	_____
____ How can you determine whether an oxidizing agent will oxidize an organic molecule to the next oxidation stage, or to an oxidation stage that is two levels higher?	Putting this information together, here is how to determine how strong a common oxidizing agent is, and how to predict the number of levels an organic molecule will be oxidized:
	_____
	_____
	_____
	_____
	Use the information you learned to create a chart showing oxidation levels and indicate the reactions that occur when an organic molecule reacts with each of the following: O_3, $KMnO_4$, OsO_4, HIO_4, H_2CrO_4, $Na_2Cr_2O_7$, CrO_3, and PCC:

Skills Check 6

Predict the products of the following redox reactions:

a.   methyl butanoate + xs LiAlH₄ followed by acid

b.   butan-1-ol + Na₂Cr₂O₇

c. *cis*-but-2-ene + H_2 with Pt

d. *cis*-but-2-ene + O_3

e. *cis*-but-2-ene + OsO_4

f. but-2-yne + Na in NH_3

g. butan-1-ol + H_2CrO_4

h. butane-2,3-diol with HIO_4

i. (R)-butan-2-ol + $KMnO_4$

j. but-2-yne + Lindlar's reagent

k. butan-1-ol + CrO_3

l. *cis*-but-2-ene + H_2CrO_4 (hot)

m. butan-1-ol + $KMnO_4$

n. but-2-yne + H_2 with Pd

o. *cis*-but-2-ene + $KMnO_4$ (hot, concentrated)

p. butanone + NaBH₄ followed by acid

q. methyl butanoate + LiAlH₄

r. butan-1-ol + PCC

s. butanone + NH₂NH₂ followed by KOH

Learning Objective 4: Know how to protect one functional group from oxidation while reacting another. (pages 518–519)

_____ What are ethane-1,2-diol and propane-1,3 diol used for?

Common protecting groups used in organic chemistry and what they are used for:

_____

_____

_____

Learning Objective 5: Be able to predict radical-based reactions of alkanes/alkyl groups. (pages 519–520)

_____ Why are reactions involving alkanes or alkyl portions of molecules radical-based?

The reason why reactions involving alkanes or alkyl portions of molecules are radical-based:

_____

_____

The mechanism for creating a halogen radical when treating Cl_2 or Br_2 with hν or heat:

	The mechanism for creating a bromine radical when using NBS treated with hν, heat, or H_2O_2:
_____ What is the mechanism for a reaction between a halogen radical and an alkane?	The mechanism for a reaction between a halogen radical and an alkane:
_____ What is the initiation phase of a radical reaction?	What the initiation phase of the radical reaction is: _____ _____
_____ What is the propagation phase of a radical reaction?	What the propagation phase of the radical reaction is: _____ _____

_____ What is the termination phase of a radical reaction?

What the termination phase of the radical reaction is: _____

_____

The probability that a given hydrogen is removed from an alkane by a halogen radical:

	1°	2°	3°
Chlorine			
Bromine			

_____ How do the relative probabilities that a given hydrogen will be removed from an alkane by a halogen radical tell you which to use for a given reaction?

How this tells you which to use depending on the type of alkane desired: _____

_____

_____

_____

Learning Objective 6: Be able to work multi-step synthesis reactions. (pages 521–524)

_____ What is the general approach for working multi-step synthesis problems when determining the product based on given reactants?

How to work multi-step synthesis problems when you need to determine the product based on given reactants:

_____

_____

Factors to keep in mind when working a multi-step synthesis reaction that requires you to determine what reactants are needed to create a given product.

1. _____

_____

_____ What are four factors to keep in mind when working a multi-step synthesis reaction where you need to determine what reactants are needed to create a given product?	Example: Work out the best method for synthesizing cyclopenta-1,3-diene from cyclopentane.
	2. _____ _____

Example: Work out the best method for synthesizing *m*-chloroethylbenzene from benzene.

3. _____

_____

Example: Work out the best method for synthesizing erythro 3-bromobutan-2-ol from but-2-yne.

4. _____

_____

Example: Work out the best method for synthesizing butanoic acid from toluene.

Learning Objective 7: Prepare for the ACS Examination in Organic Chemistry. (pages 524–527)

Your personal game plan and schedule to study for your final exam:

_____

_____

_____

_____

_____

_____

_____

_____

Integrate Skills

As a final review and practice, refer to _Preparing for Your ACS Examination in Organic Chemistry_, published by the American Chemical Society.

Apply Your Skills to New Situations

In the metabolic pathways of biochemistry, catabolic reactions are often used to oxidize molecules for energy. When oxidation involves only the loss of two hydrogens, the reaction creates $FADH_2$. Reactions that increase the number of bonds to oxygen as well as reduce the number of bonds to hydrogen create NADH. Based on this pattern, label each step that produces $FADH_2$ or NADH in each pathway shown below.

Glycolysis

TCA cycle

Beta-Oxidation of Fatty Acids

$$R-CH_2-CH_2-CH_2-\overset{\overset{\ddot{O}:}{\|}}{C}-SCoA$$

$$\downarrow$$

$$R-CH_2-CH=CH-\overset{\overset{\ddot{O}:}{\|}}{C}-SCoA$$

$$\downarrow$$

$$R-CH_2-\underset{\underset{OH}{|}}{CH}-CH_2-\overset{\overset{\ddot{O}:}{\|}}{C}-SCoA$$

$$\downarrow$$

$$R-CH_2-\overset{\overset{O}{\|}}{C}-CH_2-\overset{\overset{\ddot{O}:}{\|}}{C}-SCoA$$

$$\downarrow \quad ^-SCoA$$

$$R-CH_2-\overset{\overset{O}{\|}}{C}-SCoA \qquad H_3C-\overset{\overset{\ddot{O}:}{\|}}{C}-SCoA$$

Summary of Concepts and Analysis Methods

Summary of the Chapter:

_____

_____

_____

_____

_____

_____

_____

_____

Questions You Should Ask In Class:

_____

_____

_____

_____

_____

_____

Common Mistakes You Tend to Make but Want to Avoid in the Future:

_____

_____

_____

_____

_____

Types of Problems Needed for Targeted Practice:

_____

_____

_____

_____

_____

_____

CPSIA information can be obtained
at www.ICGtesting.com
Printed in the USA
FSHW021533100120
65815FS